WAYLAND PUBLIC LIBRARY
WITHDRAWN
3 4869 00020 7025
·WAYLAND·FREE·PUBLIC·LIBRARY·
FOUNDED 1848
INCORPORATED
WAYLAND
EAST SUDBURY 1780.
FOUNDED 1635.
1835

WITHDRAWN

D1362220

OF MICE AND MOLECULES

OF MICE AND MOLECULES

Technology and Human Survival

Eric Skjei, Ph. D.
and
M. Donald Whorton, M.D.

THE DIAL PRESS
NEW YORK

Published by
The Dial Press
1 Dag Hammarskjold Plaza
New York, New York 10017

Manufactured in the United States of America

First printing

Designed by Oksana Kushnir

Library of Congress Cataloging in Publication Data

Skjei, Eric.
Of mice and molecules.

Bibliography: p. 335
Includes index.
1. Environmental health. 2. Environmentally induced diseases. I. Whorton, M. Donald. II. Title.
RA565.S536 1983 363.7′3 82-9697
ISBN 0-385-27395-9

For

Matthew, Andrew, Mandy,
Christopher, Max, Kerry,
Jenny, John, David, Patrick,
their children,
and their children's children.

A book like this one, if it is to encourage rather than dismay, must steer a tricky course between sensationalism and sugar-coating, while trying at the same time to combine accuracy with readability. In the exacting world of science, this is never easy. Where we have erred, the faults are ours, not those of any of the individuals below who gave so generously of their time and thoughts.

Chambier Bechtel
Teddy Bricker
George Briggs
Bob Cooper
Dave Duncan
Nancy Frost
George Fuqua
Susie Jackson
Bill McNeice
Fred Prahl
Jo Ann Semones
Allen Simontacchi
Dave Spath
Ed Storey
Jim Suhrer
Ralph Thomas
Bill Thurston
Phil Woods
Laura Yoshi
Gunter Zweig

Illustration credits: Rik Olson and Sara Raffeto

CONTENTS

ILLUSTRATIONS

INTRODUCTION

Disease is part of the environment. Historically, plagues and epidemics caused by organisms in our air, water, and food have always ranked among the worst of mankind's natural enemies. Smallpox, cholera, typhus, bubonic plague—these and hundreds of other diseases have afflicted us since time began. The Old Testament records a fearful contagion called down upon Pharaoh and his people in the story of Moses' flight from Egypt. Thucydides alludes to an epidemic of what was probably smallpox in 430 B.C. that changed the course of the Peloponnesian Wars. In the Middle Ages the Black Death struck down every third inhabitant of what is now Europe, killing some 20 million people within the space of ten years.

There is nothing new about the feelings of terror that accompany environmental disease. Striking as it does at rich and poor, proud and humble, celebrated and obscure alike, it appears arbitrary and uncontrollable. In its wake comes a mood of helplessness that can sap a culture's vitality and will.

Today, in the industrialized nations of the world, most of the classic environmental diseases of the past have been conquered by medical science, aided by the development of better nutrition, better sanitation, and better health care. Only a few decades ago, for example, tuberculosis was one of the leading causes of death in the United States; in 1981 improved living conditions, new drugs, and effective screening tests have made it a thing of the past for nearly all of us. In parts of the world where these achievements are still uncommon, the fight against natural environmental diseases still rages, though today they bear exotic names like trypanosomiasis, filariasis, and schistosomiasis.

But what is new about environmental disease is that it is becoming more and more a product of our own creation. The developed nations of the world are learning that the very technology that can make human existence safer, healthier, and happier has its dark side, too.

A four-day coal smog paralyzes London in the winter of 1952. In its aftermath an excess of four thousand deaths appears in the city's mortality registers. In the United States hundreds of thousands of men and women patriotically join the nation's massive shipbuilding effort during World War II, only to learn decades later that the asbestos insulation they were exposed to in the shipyards then is a potent cause of lung cancer and other diseases. In 1972 a Japanese court finds a major chemical company guilty of dumping poisonous mercury wastes into a bay in southern Japan, and orders it to indemnify local residents for the hundreds of deaths and thousands of injuries that have resulted from consumption of mercury-contaminated fish and shellfish in the area. Throughout the 1970s Vietnam veterans exposed to Agent Orange during the war complain futilely to the Veterans Administration about a broad spectrum of health problems caused, they believe, by exposure to the dioxin-contaminated herbicide. Pesticide workers all across the country learn in 1977 that a chemical they have been injecting into the ground to kill a parasitic worm can also make them permanently sterile. In 1978 the country is rocked by the news that hundreds of families in Upstate New York will have to evacuate their homes because of evidence that toxic wastes leaking from a dump known as Love Canal may be causing miscarriages and other health problems. In 1979 the worst commercial nuclear reactor accident in history occurs on Three-Mile Island in the Susquehanna River near Harrisburg, Pennsylvania. In that same year five hundred new mothers in Michigan learn that their breast milk contains unsafe levels of PCBs, a toxic substance used in vast quantities for fifty years and now found in every corner of the globe, from Arizona desert to polar ice cap. In 1981 a government committee recommends that plutonium wastes, which will still be highly radioactive thousands of years from now, be buried in gigantic underground salt caverns in areas along the Gulf Coast.

The list of such examples is an endless and sobering one. From them we are slowly learning a very tough lesson: As the quality and the very molecular makeup of our environment become more and more a product of our own technology, we must take greater responsibility for seeing to it that the changes we cause are not toxic. In an age where we have not only the power to obliterate the earth but also to irrevocably poison the life that inhabits it, in an age where decisions made in a corporate office, research lab, or government agency thousands of miles away from us can have as intimate an impact on our health as the act of brushing our teeth, we can no longer afford to think of health as nothing more than the private concern of each individual. Whether we like it or not, we now live in a world where the well-being of our environment and ourselves is fast becoming one and the same, where health is as much a public, political matter as it is a personal, individual one, and where it is clear that if we do not act collectively now to safeguard the health of our environment, there will be less and less we can do about it in the future.

Human health transcends purely biological health because it depends on those conscious and deliberate choices by which we select our mode of life and adapt, creatively, to its experiences.

. . . To heal does not necessarily imply to cure. It can simply mean helping people to achieve a way of life compatible with their individual aspirations—to restore their freedom to make choices—even in the presence of continuing disease.

—René Dubos
("Health and Creative Adaptation," *Human Nature,* Vol. I, No. 1, January 1978, pp. 74–82)

OF MICE AND MOLECULES

1

MIXED BLESSINGS

Every advance in the human condition, it seems, brings new problems with it. Unlike the straight, steady climb of a jet plane taking off, the course of our progress is instead more like the trail of a frog climbing out of a well: great leaps upward are invariably followed by backward slippage, and long periods of retrenchment in which energy must be gathered for the next surge forward.

Throughout this century—most intensely since World War II—the industrialized nations of the world have been learning just how aptly this pattern (one might label it "frogress") applies to technological development and just how inescapable its inherent Sideslip Factor truly is. As each day brings us fresh news of the power of industrial cultures to contaminate the air, water, soil, food supply, and the very tissues of our bodies with foreign substances, many of which are highly toxic, we are forced over and over again to wonder if life on earth isn't in as much danger of being irrevocably poisoned by the substances these cultures manufacture in staggering quantities as it is of being vaporized in a thermonuclear war.

Figures on the amount of synthetic chemicals we produce each year vary, but the basic picture is this: From 1940 to 1980 production of these compounds rose from less than 500,000 metric tons to more than 8 million metric tons per year, an increase of some 1,600 percent. Of a total of some 4 million different chemical compounds now registered with the American Chemical Society, at least 65,000 are in common use, and each year about 500 or so new ones, never before known on earth, make their way into the marketplace.

Our society is so dependent on these chemicals that it could not continue to function without them. They are essential to the way we provide our food, water, clothing, housing, furniture, appliances, drugs, and much more—to everything from the way we fight forest fires to how we keep our plumbing from clogging up.

But one in every three of these compounds has some toxic properties. Some inevitably escape into the ambient environment in the process of being created, used, and disposed of. Many are extremely resistant to biodegradation. And in general we know far too little about how the substances that we introduce into the environment affect living organisms. Between the sciences of environmental chemistry and biology there is a large and dangerous knowledge gap.

The power of the synthetic substances we manufacture to disperse throughout the biosphere is awesome. Before they were banned in 1978, enough fluorocarbons were released from spray-can propellants to accumulate in the stratosphere at such levels that they now threaten to deplete the ozone layer, which is all that protects us from carcinogenic levels of exposure to the sun's ultraviolet rays. Fluorocarbons are so durable that we will probably not be able to assess the true scope of this potential hazard until well into the next century.

PCBs, another synthetic substance widely used for half a century, primarily to insulate electrical equipment, until they were banned in 1979, have been so widely distributed in the environment that they have been detected in places as remote as five meters under the polar ice cap, in the middle of the Sargasso Sea, and in the most isolated areas of the Arizona desert. Most of us now carry some PCBs in the tissues of our own bodies.

In the water toxic substances have been turning up with alarming regularity in the United States in the last few years, particularly in groundwater aquifers all across the country, often as a result of leakage from nearby waste dumps. Unfortunately, an aquifer cannot be easily protected from this type of pollution, and once polluted it may stay that way for decades. The only safety measure that can be taken in many cases when well water drawn from an aquifer is found to be dangerously polluted is to close it to human use, and in many parts of the country loss of these precious resources is coming at a time when our need for water is beginning to outstrip available supplies. In some of these cases recycling used water is the answer, but in others it is not, since recycling can cause some pollutants to become concentrated to even higher levels in the water.

One of the most ironic examples of the Sideslip Factor at work is chlorination. Early in this century cholera, typhoid fever, and numerous other waterborne infectious diseases were a major public health problem in this country. Then it was discovered that adding a small amount of chlorine to water supplies would kill disease-causing bacteria without apparent ill effect in human users. Soon chlorination had become a universal water treatment practice.

Then in the 1970s scientists discovered that when chlorine was mixed with certain natural organic substances, such as fulvic and humic acids released from topsoils, a new type of compound, called trihalomethanes, was often formed. Trihalomethanes, it turns out, are carcinogenic in laboratory animals.

One of the most significant uses of synthetic chemicals in industrialized cultures is in fertilizers and pesticides. Both uses, we have been learning, can have side effects that are severely detrimental to the environment and the life it supports. Runoff from fertilized fields can carry toxic nitrogen derivatives into local water supplies, causing a blood disease in infants known as "blue baby." The nutrients in the runoff can also cause excessive algae growth in surface waters, which in turn can prevent the sun from penetrating below the surface. As a result photosynthesis in other aquatic flora stops, killing both them and all the other organisms in the aquatic food chain that depend on them.

Chemical pesticides raise particularly complex issues. Insects, weeds, and other pests destroy something like half of the world's food crops each year; without pesticides, that figure would be much higher. But since pesticides are expressly designed to kill living organisms, their potential to cause toxic pollution of the biosphere is much greater than that of many other synthetic compounds. Ever since 1962, when Rachel Carson drew our attention to the hazards of DDT in *Silent Spring,* we have learned to be much more careful about how we use pesticides and how we test them for toxic effects before we use them. But accidents and oversights still occur as a result of which people are seriously harmed by pesticides we thought were safe.

In the next two decades, as the world's population climbs from its current 4 billion to over 6 billion, and the amount of arable land barely increases at all, the pressure to use even greater quantities of pesticides will become intense, especially in the Third World countries where virtually all of this population explosion will take place. The grim irony here is that because these are also frequently the nations where starvation is most rampant, these pesticides may become even more toxic, since a diet deficient in protein tends to make an individual even more susceptible to many of them. In general, the worst problems with environmental disease in the future will probably be seen in these countries, since their struggle to rapidly make the transition out of an agrarian and into an industrialized culture will mean that they will become exposed to the unique health hazards that accompany technology before they have truly conquered the traditional infectious diseases that they suffer from now.

As well as synthetic pollutants, industrial societies tend to contaminate the biosphere with the byproducts, emissions, wastes, and residues of the natural resources that they consume in such great quantities, and many of these are also highly persistent and potentially toxic. For instance, the burning of enormous amounts of fossil fuels—oil, coal, gasoline—is a major characteristic of industrial nations, one that can have a profoundly toxic effect on the atmosphere in localized areas unless it is carefully controlled. Pollutants like sulfur dioxide, nitrogen dioxide, carbon monoxide, ozone, and others emitted by the factories, power plants, and motor vehicles that burn fossil fuels have a clear potential for harmful health effects.

Asbestos is another good example of a natural substance that is now found in the air, water, in some foods, and in the lungs of virtually every city dweller in this country. Because thousands of industrial applications have been found for it over the last century, billions of kilograms of asbestos have been mined, milled, processed, used, and eventually discarded in dumps and waterways, and there has been a steady release of breathable and highly hazardous asbestos fibers into the environment at large.

There could hardly be a more "natural" substance than carbon dioxide, which every human being emits into the biosphere every time he or she exhales. However, carbon dioxide is yet another one of the innumerable byproducts of an industrial society that may cause environmental problems. Any kind of combustion in any kind of society releases some carbon dioxide into the air, but the enormous amount of combustion that takes place in industrialized societies, particularly of fossil fuels, causes the emission of vast volumes of carbon dioxide into the atmosphere, far more than can be disposed of by such natural absorbers as plants, which take in carbon dioxide and give off oxygen. As a consequence carbon dioxide levels in the atmosphere have been slowly but steadily rising over the last century. In the atmosphere, carbon dioxide may trap the sun's heat, preventing it from escaping back out into space, causing the atmosphere to function more and more like a giant greenhouse canopy, and possibly causing the global temperature to rise. The consequences of even a very small increase in the earth's temperature—and this is still an extremely controversial issue—would be catastrophic: melting ice caps would lead to rising sea levels and severe coastal flooding all over the world; climatic patterns would change, turning our most productive farmlands into deserts, bringing famine and massive population shifts.

Another painful lesson we're learning is that a polluting society also tends to be a wasteful society. Each year in the United States alone we create staggering amounts of waste, and some 60 million metric tons of it—mostly generated by business and industry—is toxic. Until quite recently much of this toxic waste was simply drained into natural waterways, dumped out on the ground, or dropped into "landfills" that were often little more than holes in the ground. Most of it was promptly forgotten about, and much of it, we now know, soon began to leak out of its containers and sink into the nearby soil. Ironically, much of this waste is reusable—one company's wastes, we're learning, may be another company's raw materials. Despite the passage of recent laws intended to track toxic wastes from the moment they are generated to the moment they're disposed of, and to assure that they are disposed of safely, we will probably be uncovering older, forgotten dumps—and dealing with their health hazards—for years to come.

Many of the problems that industrialized societies face with environmental health hazards are epitomized by their attempts to harness the atom for peaceful purposes, and this applies to the problem of toxic wastes, too. The seventy-five-odd commercial nuclear reactors dotted around the country produce a radioactive waste, plutonium, that is one of the most toxic substances known to mankind—certainly it is the *only* one we know of that will be hazardous half a million years from now. As of this writing, our plutonium wastes were steadily piling up in special, crowded storage depots because we still hadn't developed a safe way to dispose of them.

In this book we discuss some of the major pollutants and their health effects that pervade our air, water, and food. We also discuss hazardous wastes and radioactivity. We describe the problems these things pose, but we also try to include the bright side, the side that does not get told in the daily news reports, the fact that the same human ingenuity that created most of these problems in the first place is hard at work on ways to resolve most of them. Throughout this book, where it is appropriate, we touch on some of these solutions.

To a certain extent, dividing the biosphere into compartments like air, water, and so forth, as we tend to do in this book, makes sense, since there are characteristic pollutants in each one that aren't found at all or at any level of significance in the others. There are gaseous pollutants in the atmosphere that do not appear in the soil or water. There are liquid pollutants in water supplies that are not found to a significant extent in the soil or the air. But despite this conventional and convenient compartmentalization of the environment, the reader should bear in mind throughout this book that the most fundamental principle of environmental matters is *interdependence.* There is always a great deal of exchange and interaction between these various zones. Gases emitted into the air may mix with its moisture, become liquid droplets, wash to earth in rain or snow, and enter into the soil, into the plants that grow into it, into the bodies of animals that eat these plants, and into the bodies of the animals that eat those animals. For example, this is roughly what happens to radioactivity from nuclear weapons testing conducted in the atmosphere. Or, for another example, pesticides initially applied to the soil may be carried by storm runoff into local waterways and from there may, through evaporation, be drawn up into the atmosphere by the sun's heat, along with the billions of kilograms of water lifted up from the earth's surface every day by the same natural process. Nothing, it should be kept in mind, is permanently confined to any one compartment of the environment.

This interdependence has important political implications. Although we focus in this book on environmental health issues in the United States, the issue is already a global one. Radioactivity from nuclear testing in China drifts eastward, passing across the Pacific and over North America. The fluorocarbons that we mentioned earlier, thought to be depleting the ozone layer above our heads, disperse rapidly around the globe no matter where they first originate. Sulfur and nitrogen oxides from power plants in Great Britain have for decades been wafted by wind currents across the North Sea to Northern Europe, especially Scandinavia, where they contribute to acid rainfall. Hundreds of lakes in Sweden and Norway have been depleted of life by this process, and it may also be interfering with the growth of pine trees in a variety of complex ways. The United States and Canada were just beginning to come to grips with this problem as this book was being written. For some years, acid rain from American and, to a lesser extent, Canadian power plants has been crisscrossing the border between New England and southeastern Canada, causing the same kinds of damage: barren lakes and complex effects on forest growth.

Environmental pollution defies abstract territorial boundaries. Pesticides sprayed on one farmer's crops can be carried in the air or water to a neighbor's fields, or can run off into surface and groundwater supplies that provide drinking water to people living hundreds of miles away. A number of pesticides banned in this country, including DDT, are routinely used on crops grown elsewhere, many of which are then shipped back here for sale in our markets. FDA spot checks at the border for illegal residues of these pesticides cannot possibly be expected to prevent them from ever making their way back into the United States.

Understanding the interdependence of the environment is essential to understanding the impact of our technology on it and thus on us. It is also essential to an accurate appreciation of the true costs that we must pay for technological progress, and the only way we can begin to learn how to make intelligent decisions about the tradeoffs that we are willing to make to allow further development to take place. Ultimately, as Lewis Thomas has pointed out, understanding interdependence may be the only way to perceive correctly our own role on this planet:

> The oldest, easiest-to-swallow idea was that the earth was man's personal property, a combination of garden, zoo, bank vault, and energy source, placed at our disposal to be consumed, ornamented, or pulled apart as we wished. The betterment of mankind was, as we understood it, the whole point of the thing. Mastery over nature, mystery and all, was a moral duty and a social obligation.
>
> In the last few years, we were wrenched away from this way of looking at it, and arrived at something like general agreement that we had it wrong. We still argue the details, but it is conceded almost everywhere that we are not the masters of nature that we thought ourselves; we are as dependent on the rest of life as are the leaves or midges or fish. We are part of the system. One way to put it is that the earth is a loosely formed, spherical organism, with all its working parts linked in symbiosis. We are, in this view, neither owners nor operators; at best, we might see ourselves as motile tissue specialized for receiving information—perhaps, in the best of all possible worlds, functioning as a nervous system for the whole being.[1]

Like any other specialized field, environmental health has its own language. Some of the terms we use often in this book may need a few words of introduction.

Toxic, a central concept, simply means capable of causing illness. The types of illnesses caused by environmental toxins are conventionally divided into acute and chronic. Acute illnesses are those which appear soon after exposure to a toxic compound, last for a relatively short time, and then resolve themselves, even if the resolution is in death. The term *subacute* is also occasionally used to describe disorders with subtle symptoms that are not immediately obvious without special tests. Lead workers, for example, often appear to be much healthier than a thorough medical examination reveals them to be. *Chronic* illnesses, by contrast, are those that may appear years or even decades after exposure, and which may "smolder," unresolved, for the victim's lifetime.

There are three special kinds of toxic hazards that have special relevance to environmental health: *carcinogens, mutagens,* and *teratogens.* As most of us know, a carcinogenic substance is one that causes cancer. A mutagenic substance is one that causes changes in the genetic material of a cell. Spontaneous, natural mutations occur in our body cells all the time; the vast majority of them cause no damage, and even when they do it is usually limited to the lifetime of the cell they occur in. But in rare cases the cell may continue to grow and divide after a mutagen has altered its basic genetic structure, and if this mutation is passed on to succeeding generations via egg or sperm, it may cause birth defects, inherited diseases, mental deficiency, increased susceptibility to disease, and a host of other abnormalities and disorders. If the mutated genes are recessive, it may take more than one generation for these effects to show up. Mutagenicity and carcinogenicity are related in some way, but we're not yet sure just how. Radiation is probably the best-known example of a mutagenic environmental hazard.

Over the last few decades a number of sophisticated tests have been developed to help predict the mutagenic potential of the chemicals and other substances we use so heavily, and screening for mutagenicity is now a routine part of most regulatory programs aimed at controlling potential environmental hazards.

Teratogenic literally means "monster-causing." A substance is teratogenic if it can cause birth defects, as many mutagens can.

Another central concept in this book is *contamination,* which, despite its ominous sound, simply denotes the presence of a specific substance, toxic or not, in the biosphere. Contaminants that are clearly dangerous or undesirable are also known as *pollutants.*

Contamination is measured in terms of the *concentration* of a substance in the environment, and there are a number of different conventions governing the measurement of concentration. The most common system makes use of metric units, particularly the milligram (one-thousandth of a gram, abbreviated *mg*) and the microgram (one-millionth of a gram, abbreviated *µg*). Occasionally, in very refined, ultrasensitive measurements, nanograms (one-billionth of a gram), picograms (one-trillionth of a gram), and even smaller units may be used.

Obviously, a contaminant concentration of one milligram per kilogram is the same as one part of contaminant per million parts of noncontaminant, or 1 ppm, and this alternative method of indicating relative concentrations is also widely used, particularly for air pollutants, food additives, and pesticide residues.* The terms *ppm* and *ppb* (parts per billion) are common; *ppt* (parts per trillion) appears only rarely.

Exposure is a way of saying that the contamination in the environment has passed into an organism; a human being is exposed to a toxic compound if some amount of it has entered his or her body. Exposure does *not* mean that a person has merely been in the proximity of a toxic substance. For example, if you walk past a drum bearing a warning label and containing a toxin, you are not necessarily exposed to whatever it contains. But if the drum leaks its contents into the air or soil, and pollutes the air you breathe or the water you drink, you probably will be exposed to its contents.

Like illnesses, exposures may also be subdivided into acute and chronic types. Acute exposures are those that occur over short periods of time, often to high concentrations of a hazardous substance. Chronic exposures, which are much more common among the general public, involve longer periods of time and, for the most part, lower concentrations.

Dose is the term for measuring exposures. Basically, the dose a person exposed to a toxic substance receives is dependent on its concentration in the immediate environment and the duration of the exposure. However, because the interaction of human beings and the environment is a complex, constantly changing process, numerous other factors may also play a part in determining dose. Dose can

* Parts per million can be converted into a percentage by dividing by 10,000; 10,000 ppm equals a 1 percent contaminant concentration.

be a function of weather conditions, the persistence and solubility of the toxic substance in the biosphere, the size of its molecules or particulates, the presence of other compounds in the environment that it may react with, the age and overall health of the exposed individual, whether the substance is inhaled, swallowed, or absorbed through the skin, and the effectiveness of the body's natural defenses in detoxifying the substance and eliminating it from the body. Some substances, like asbestos, become virtually permanent contaminants in the body once they have penetrated far enough into the lungs or other organs. Others, such as methanol, are metabolized and excreted from the body in a matter of hours at most.

The tendency for some substances to collect in the tissues of a living organism and stay there is known as *bioaccumulation.* Their tendency to move up the food chain as one species consumes another, becoming ever more concentrated as they go, is called *biomagnification.*

Traditionally a *threshold* was a measurable level of exposure to a toxic substance below which there would probably be no adverse health effect and above which there probably would be. The setting of safety standards for the work site as well as the general environment often involves the tacit assumption that approximate thresholds can be determined, monitored, and enforced for the toxin in question. But this assumption has been subjected to various criticisms in the past few decades. First, it is often pointed out that, whether we can measure it or not, it is entirely possible that *every molecule* of every substance we take into our bodies has some effect on us. It may not be a detectable effect and it may not be harmful or long lasting, but it is an effect. Thus, this argument goes, the concept of a specific cutoff point below which a substance is treated as though it didn't exist and above which it is considered harmful is misleading. Far more appropriate, proponents of this view argue, is the assumption that these substances have a *range* of effects, beginning at one end with those that are imperceptibly molecular and extending to the catastrophically toxic, ultimately fatal effect at the other end of the spectrum.

The two main disciplines involved with the estimation of the toxicity of environmental hazards are *toxicology* and *epidemiology.* Toxicologists are the ones who use all those rats and mice, as well as rabbits, dogs, monkeys, and a few other animals. Generally, toxicologists attempt to predict toxic effects. When new chemicals are tested, for example, they are fed to lab animals, or painted on their skins, or infused into the air they breathe, at various concentrations. Some of the animals die or are killed during the test and are carefully examined for signs of toxic effects. Eventually all are sacrificed and their tissues are carefully analyzed. The estimated acute toxicity of the substance being tested is then expressed as its LD_{50}, a term that stands for "lethal dose, 50 percent," meaning the amount of the substance that will cause the deaths of one-half of a group of test animals.

An LD_{50} is customarily expressed in terms of milligrams per kilogram of body weight. Thus, a substance whose LD_{50} is 2 mg/kg is five times as toxic as one whose LD_{50} is 10 mg/kg. In general, substances with LD_{50} values below 50 mg/kg are considered highly toxic. Those with values between 50 and 500 mg/kg are considered moderately toxic, and those with values above 500 mg/kg are regarded as least toxic.

Toxicology studies using animals are known as *bioassays,* and there are dozens of different kinds of them. Some take only thirty days or less to conduct, and test the effects of acute exposures. Others take much longer, as much as four years in some cases, and test for the effects of chronic exposures. Still other toxicology tests are specifically designed to focus on a special type of toxic effect, such as teratogenicity. All, unfortunately, are quite expensive, costing anywhere from two hundred thousand to five hundred thousand dollars to complete.

A typical chronic-effect bioassay works this way: A large number of animals (usually two hundred to five hundred) of both sexes are fed the substance under investigation at different dosage levels for their entire lives, which is about two years for rats. Different dosage levels are fed to different groups throughout the test. Usually one will be a relatively low dose, one a moderate dose, and one a high but not lethal dose (otherwise this whole group would die before the assay was over). Controls are kept under conditions identical to those of the test animals but are not fed any of the substance being tested.

Throughout the entire experiment these groups of animals are kept under careful observation and a number of different variables—behavior, weight change, blood chemistry, organ changes—are regularly monitored or measured. Periodically some animals from each group are sacrificed, dissected, and carefully examined to check for abnormalities—tumors, lesions, and so forth. At the end of the study the entire colony is sacrificed and tissue sections are taken and scrutinized for irregularities.

Ideally what is found is a clear relationship between increasing doses and increasing illness: animals exposed to higher doses show more tumors; animals exposed to lower doses develop fewer. Such a finding strongly suggests that there is a direct relationship between the amount of the compound taken into the body, the dose, and the degree of disease resulting, leading to the logical conclusion that the way to prevent the substance from causing disease is to keep exposures to it below a certain level.

If such a relationship does appear, the next step is to try to calculate the "no-effect level" (NOEL), that is, the amount of the substance a test animal can absorb for its entire lifetime with no ill effects, or at least with no more ill effects than nonexposed controls develop. (There is always some disease, even in the controls.)

If the NOEL turns out to be something like 1 mg/kg (remember that's equal to 1 ppm), the next step is to figure out the human equivalent. Using a figure of 70 kilograms (158 pounds) for the weight of the average person, the first assumption to make is that if the NOEL in humans is the same as in rats, then a human being can probably take in 70 milligrams of the substance over his or her lifetime without being harmed. This part of the NOEL calculation, known as the "species-to-species extrapolation," is especially tricky because not all animals react to comparable doses of the same substances in the same ways. For example, human beings are one hundred times *more* sensitive to the tranquilizer Thalidomide than rats are. Man, mice, and rabbits are generally more susceptible to cancers caused by skin absorption than rats are. And even substances with a similar degree of toxicity in different species may not attack the same organ systems. Rats exposed to high levels of the carcinogen benzidine tend to develop tumors in their intestines; human beings exposed to hazardous doses of benzidine are more likely to develop urinary bladder cancer.

So, since human beings are not rats, a safety factor of 10 is figured into the NOEL at this stage, bringing our hypothetical example above down to 7 milligrams. Then, since there are special subgroups in any population—old people, very young people, those who are already ill—who are much more susceptible than the average, reasonably healthy adult to environmental disease, a second safety factor of 10 is again figured in, bringing the NOEL down to a very low 0.7 mg. And this will be the figure that is used in establishing levels of acceptable exposure to this substance.

As crude and arbitrary as it may seem in many ways, the bioassay process tends if anything to err on the side of caution and safety. And so far, in terms of direct tests for toxicological effects, it is the best thing we have.

There is a very good reason for what often seem to be the absurdly high dosage levels used in these tests. Dr. Samuel S. Epstein explains it well:

> Let us suppose that humans and rats are equally sensitive to some chemical carcinogen which causes one case of cancer in every 10,000 persons (or rats) to which it is given. If 220,000,000 Americans were exposed to this chemical, 22,000 cases of cancer would occur. On the other hand, if fed to the typical fifty rats used in an experiment, the chances that even one rat would get cancer is one-half of one percent; 10,000 rats would have to be fed the chemical (at human dosages) to observe even one cancer.
>
> Given that a human-level dosage might not produce a detectable result in a small rat population, even for a carcinogen which may be the cause of many thousands of cases in humans, the only alternative is to increase the dose, and thereby increase the probability of inducing a cancer in a particular animal. Once the dosage is increased above real-life levels, scientists must begin to make assumptions about their findings in the following vein. If the equivalent of, let us say, 1,000 cans of saccharin-containing soda per day produces one tumor in 50 rats, then it is plausible to assume that 100 cans will produce one-tenth as many tumors, or one tumor in 500 rats. Continuing the argument, the equivalent of 10 cans per day would produce one-tenth as many as before, that is, one tumor in 5,000 rats. Finally, one can per day would produce one tumor in 50,000 rats. If, as we assume, humans and rats are equally sensitive to this carcinogen, one can of soda per day would produce one tumor in 50,000 people, or about 4,000 cases of cancer in the nation's population.*[2]

The argument is sometimes made, by those who distrust the bioassay method, that the massive doses used can themselves cause tumors. This is not true. A substance that is not carcinogenic will not induce excess tumors in test animals no matter what dose is used. The critical difference between carcinogenic and noncarcinogenic substances is a qualitative rather than quantitative one. However, massive doses of any substance, even water, administered to a test

* The assumption that this relationship between dose and cancer incidence is a linear one is not necessarily true; in other words, at low doses the corresponding incidence of cancer may be even lower than Dr. Epstein suggests here.

animal may so overwhelm its defense mechanisms that it dies of other causes.

Epidemiology is the study of the actual effects of substances on people exposed to them. Industrial workers, for example, are often exposed to much higher concentrations of potentially hazardous substances than are other groups, so studies of the effects of these substances on their health can often give us clues about how hazardous they may be in the general environment. In a sense, every individual in a highly industrialized country is an unstudied subject in a national epidemiology study.

The concepts of *organic* and *inorganic* are also important. Inorganic refers to nonliving matter, like asbestos, or lead. The term "organic" includes but is not limited to living matter. At its broadest, it means anything containing carbon, one of the basic building blocks of life. Since coal, oil, and diamonds contain carbon, they are all technically organic. When chemists in the nineteenth century discovered that it was possible to use coal and petroleum derivatives to make artificial substitutes for many natural compounds, they created the field of organic chemistry, thus making possible such apparently self-contradictory terms as "synthetic organic."

Finally, it is important to understand the concept of *synergy.* Of the thousands of substances emitted into the air, water, and soil by industrial societies, many are stable and inert, but many others are not. In the presence of sunlight, water, air, heat, and some of the thousands of other compounds, both natural and artificial, that permeate the biosphere, they may undergo complex chemical transformations, becoming in some cases even more toxic than they were before. For example, under certain atmospheric conditions the pesticide parathion may be converted into a compound known as paroxon, which is ten times as toxic as its progenitor. In cases where the effect of substances in combination is greater than their separate effects would be if they were simply added together—when, in other words, the whole is greater than the sum of its parts—the correct term for this phenomenon is synergy.

Synergy is not necessarily evil. It can also have positive applications. For instance, one of our most benign pesticides, a natural plant extract known as pyrethrum, has traditionally been made more effective by the addition of a nontoxic synergist, piperonyl oxide. This has a double effect: it makes a relatively small amount of pyrethrum go a lot farther, in terms of its impact on pests, and it keeps costs down since the synergist is less expensive than pyrethrum extract.

The realization that chemicals that can freely interact in the biosphere may become more toxic than before is hardly new. In 1962 Rachel Carson described a case in which "chlorides, chlorates, salts of phosphoric acid, fluorides, and arsenic" were collected in holding ponds from 1943 to 1951 by the Army Chemical Corps at the Rocky Mountain Arsenal in Colorado, near Denver. As early as the late 1940s, complaints of crop damage, livestock disease, and occasional human illnesses of a mysterious nature began showing up among farmers in the region, but until a good ten years later no one connected these disorders with the Army's holding ponds.

Then in 1959 a study of well water on a number of these farms determined that the groundwater they tapped into had not only been contaminated by many of the substances held in the Army's ponds but also that those substances had somehow recombined to form 2,4-D, a widely used, moderately toxic herbicide. As Carson observes,

> It had been formed there from other substances discharged from the arsenal; in the presence of air, water, and sunlight, and without the intervention of human chemists, the holding ponds had become chemical laboratories for the production of a new chemical—a chemical fatally damaging to much of the plant life it touched.[3]

In much the same way, chemical reactions *inside* the body can also transform relatively innocuous substances into other compounds, known as metabolites, most of which are harmless but some of which can be deadly. Methanol, for example, becomes hazardous when ingested because the body converts it into formaldehyde. Some causes of cancer, such as benzene, appear to become carcinogenic only after they have been metabolized in the body.

Highly technological societies know a great deal about many of the separate substances, natural and artificial, that make their existence possible. In some cases they also know a great deal about the acute toxicity of many of these same substances. Much less is known, in general, about their chronic toxicity. What we know least about, and must clearly begin to pay much closer attention to, is what happens to these substances once they are liberated into the biosphere, and become free to evolve in ways that we cannot even keep track of, let alone control.

2

THE BODY ENVIRONMENTAL

Without air a human being suffers irreversible brain damage in five minutes. Without water we cannot survive more than a few days, perhaps a week in the best of circumstances. Without the vitamins, minerals, carbohydrates, fats, and proteins that food provides, our metabolic fires slowly die down, beginning to consume our own tissues after a month of so of starvation, and eventually stop altogether. The interdependence of body and biosphere is clearly a vital one. Since we are by nature organisms that must draw elements of the world around us into our bodies in order to survive, the quality of the environment has profound implications for our well-being.

In this critical interaction the role the body plays is not that of a hapless victim. Rather than passively accepting and incorporating anything and everything that happens to be present in its environment, the body actively screens what it encounters and selects out what it needs, at the same time carefully deflecting many of the pollutants that have the potential to harm it. Even those hazards that do manage to find their way into the body are confronted by an impressive array of defense systems. The best model for the way the body functions in its environment is that of a membrane.

A membrane is simply a barrier that allows different concentrations of substances on either side of it. If, for example, one adds common table salt to a tank of water, the salt disperses naturally throughout the tank until its concentration is the same everywhere. But if a membrane permeable to water but not to salt is then placed across the middle of a second tank and the same amount of salt is

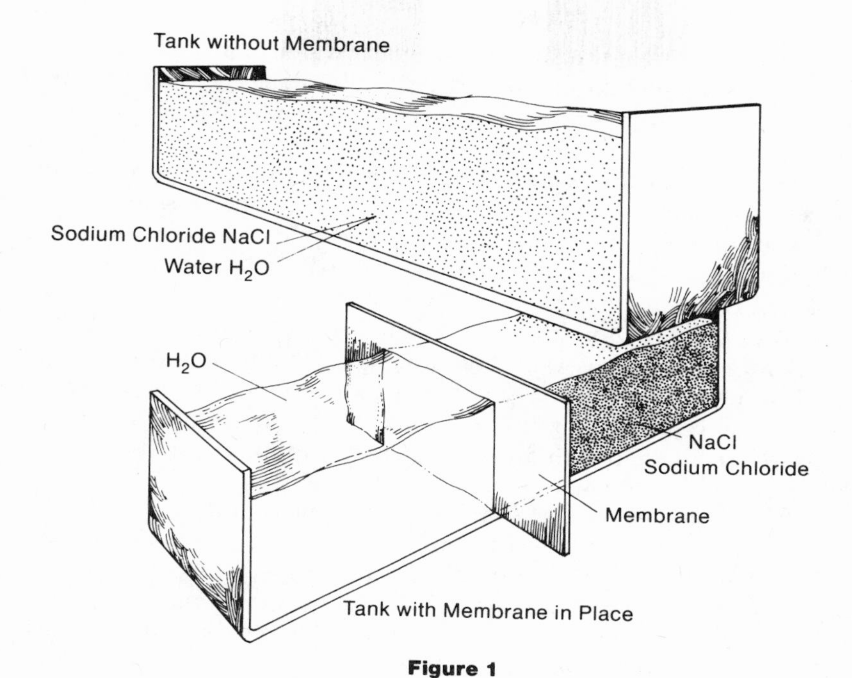

Figure 1

then added to one side, it will not diffuse across the membrane and the salt concentration will be twice as high on one side of the membrane as it was in the first tank.

Membranes are, in short, a way of preserving a state of disequilibrium. Strange as it may sound, the life of the body is fundamentally a matter of attaining and preserving highly dynamic states of disequilibrium, and membranes characterize this process on every scale from the smallest cell to organs as large as the skin. They maintain high levels of water and other fluids inside the body, which must be kept far wetter than its terrestrial surroundings. They help preserve a constant body temperature in the face of external temperature variations that can range from far below zero to well above 37°C. They actively scavenge the oxygen and nutrients out of the biosphere that we must have to exist, while simultaneously helping protect us against the harmful organisms and toxins that pervade our air, water, and food.

For example, in each of the 50 trillion or so cells that make up the human body, water and other fluids pass freely across the membrane that constitutes the cell wall, carrying nutrients in and wastes out. In the case of water the cell wall is permeable in both directions.

In fact, despite the astonishing amounts of water that pour back and forth across the membranes of any cell wall, the net volume of water inside the cell is always the same—a remarkable physiological balancing act.

Other substances, especially those critical to the metabolism of the cell, are allowed in just as easily as water, but once inside the cell they are not allowed to leave quite so easily. The concentration of potassium in the blood, for example, is ordinarily about 4 mEq*; inside the cell wall, however, it is typically about 140 mEq. With other substances this situation is reversed—they are actively kept out of the cell by its membranous wall. Sodium, for example, is found in the blood at levels like 140 mEq, but inside the cell its concentration is normally more like 4–5 mEq. Oxygen is freely allowed to enter cells, but is then combined with other substances and prevented from slipping right back out again.

On a larger scale the skin itself, our most visible membrane, exhibits the same characteristics as any other in the body: highly selective absorption plus equally discriminating rejection of potentially harmful substances. To take just one example, most of the ultraviolet light we are exposed to is deflected by our skin, through its production of a substance called melanin, which gives skin (and the irises of our eyes) its various hues. But at the same time the skin allows a small amount of sunlight to enter and interact with a specialized alcohol it produces in order to create the vitamin D that our bones must have if they are to develop normally when we are young and remain strong and healthy in adulthood.

In the crucial interaction between the human organism and the ambient environment, three organ systems—all functioning essentially as membranes—serve as the major points of contact between body and biosphere. These three "routes of entry," as they are also known, are the respiratory tract, the gastrointestinal tract, and the skin.

THE RESPIRATORY TRACT

Along with carbon and hydrogen, oxygen is essential to the body's metabolic engines, and it is the function of the respiratory tract to provide a constant supply of warm, moist, clean air to the

*"mEq" is the abbreviation for milliEquivalent, a way of measuring and comparing electrically charged substances dissolved in the blood that accounts for their differences in molecular weight.

membranes deep in the lung that steadily infuse it into the bloodstream.

From our first cry at birth to our final expiration, nothing is more important or more routine to us than our breath. Throughout life, twelve to sixteen times per minute while we're at rest, the diaphragm rises and falls, the muscles of the rib cage contract and expand, and we breathe. Normally each breath brings about half a liter of air into the body, but during hard physical labor or exercise that volume can rise sharply, to as much as 6.5 liters per inhalation in a young, well-trained athlete.

From the moment air is drawn in until it is transfused across the membrane of the lungs and into the blood, it is extensively cleaned and conditioned by a series of protective mechanisms in the respiratory tract. In the upper respiratory tract, which comprises the nasal and sinus passages, the throat, and the larynx (voice box), incoming air is first crudely screened by nostril hairs for particulate matter above a certain size (10 microns in diameter, or about 1/2,500 inch). Smaller particles are unaffected by this initial screening process. At the same time, contact with the moist spaces inside the nose warms and humidifies it, preparing it to release some of its oxygen content into the bloodstream. (Breathing through the mouth diminishes the effectiveness of these natural purifying and filtering processes.)

Highly soluble gases will tend to begin dissolving as soon as they come into contact with the moisture present throughout the respiratory tract. In general this process helps prevent gas molecules from penetrating deeply into the lungs, but the side effects can themselves be unpleasant and even hazardous. For example, sulfur dioxide (SO_2) readily interacts with water to form sulfuric acid. Since droplets of liquid sulfuric acid are heavier than molecules of gaseous SO_2, they tend to become deposited on the walls of the nasal passages and throat. The irritating effect of the deposition of a caustic acid on these surfaces is annoying and potentially harmful, since it can cause serious burns, but it also tends to have a beneficial consequence: The irritation prompts an individual exposed to SO_2 to breathe more shallowly and to leave the environment where the exposure is occurring. Thus, though the irritation may itself be somewhat distressing, it often functions as a warning signal and prevents more dangerous exposures deeper in the respiratory tract. Gases that are not so highly soluble, such as nitrogen dioxide, are of course unaffected by the water vapor present in the upper respiratory tract.

Gases of any kind, soluble or not, may combine with microscopic particles of solid or liquid matter in the air in a process known as *ad*sorption (not the same as *ab*sorption), a kind of mechanical synergy in which gas molecules become physically attached to particles of solids or liquids. If the gas happens to be a fairly soluble one, like SO_2, adsorption on particulates may reduce its propensity to interact with water molecules and thus enhance its likelihood of traveling deeper into the respiratory tract than it normally would. For this reason SO_2 and other gases are much more hazardous in dusty, particulate-laden air than in clean, highly filtered air. As we explain in Chapter 4, one of the most notorious air pollution disasters of this century was probably caused by the adsorption of SO_2 on the particulates present in coal smoke.

The nose and throat are lined with special cells that emit mucus, that thick, sticky substance most often regarded as a nuisance but which in fact serves a very useful function. The nasal passages are so convoluted that incoming air is forced to make numerous sharp, sudden changes in direction as it is drawn into the body. Larger, heavier particulates tend to overshoot these directional shifts and to collide with the walls of the nasal passages. Like flypaper, the presence of a coating of sticky mucus on these surfaces assures that a sizable percentage of the particulate matter that hits them will stay there, to be expelled later by sneezing or nose-blowing. In fact, the "runny" nose that can be such an unpleasant part of the common cold is a sign that the body has naturally stepped up its nasal mucous secretion because it has detected invasion by the bacteria and viruses responsible for colds and flu.

In addition to the humidity and sticky surfaces of the upper respiratory tract, much of the lower respiratory tract, in particular the throat and the passages just below it, is equipped with yet another ingenious defense system, one composed of millions of tiny hairlike growths called cilia. Coated with a layer of mucus and rapidly undulating in uniform waves at a rate of about one thousand sweeps per minute, the "ciliary escalator," as it is commonly called, in effect *combs* incoming air for particulates that can be trapped and then pushes them back out again the way they came in, ultimately to be either coughed up or swallowed. (In the latter case, pollutants that could have posed lung problems may eventually wind up causing problems in the GI tract instead.)

The ciliary escalator is capable of extremely effective clearing of inhaled air. At least one study found that it could remove over half the total amount of particles over 5 microns in diameter inhaled in

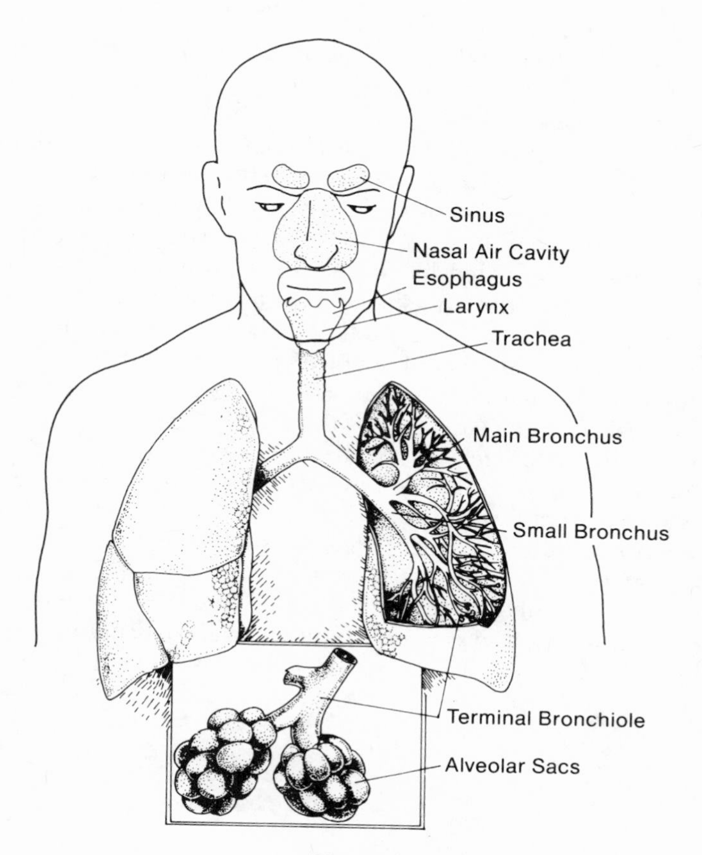

Figure 2

a twenty-four-hour period. One reason cigarette smoking is so hazardous is that it not only introduces high levels of toxic gases and particulate matter into the lungs, but also tends to paralyze the ciliary escalator, depriving the respiratory tract of one of its main lines of defense.

From the larynx down, the respiratory tract bears an uncanny resemblance to an upside-down shrub, and in fact is often referred to as the "bronchial tree" or "pulmonary tree."

As air travels downward along the pulmonary tree, the passageways that carry it become smaller and smaller, from trachea to bronchi to bronchioles, until finally they end in tiny clusters of tissue and capillary shaped something like popcorn balls or clusters of grapes and called "alveoli" (the "leaves" of the pulmonary tree). It is in these lower passageways, especially in the alveoli, that the business of transfusing oxygen into the bloodstream and withdrawing waste carbon dioxide out of it takes place.

The lower respiratory tract comprises two equal-sized lungs, which fill the chest. Although they are often depicted as hollow, balloonlike organs, the lungs are in fact much more like enormous sponges, crammed as they are with some 300 million alveoli. The vast surface area provided by the alveoli greatly increases the gas-exchange capacity of the lungs over what it would be if they were in fact nothing more than large, hollow balloons. In fact, the nearly one hundred square meters of surface area in the lungs make them by far the largest organ in the body, much larger than the two square meters of skin on the surface of the average human body. This amount of gas-exchange surface area is quite a bit more than we need for sheer survival. As is true of most of our organs, nature appears to have been working with a sizable safety factor when she designed the respiratory tract, since a human being can get along quite well with only one lung.

By the time inhaled air has reached the alveoli, it has been warmed, moistened, and cleared of many of its particulate pollutants and possibly also some of its gaseous pollutants, depending on their

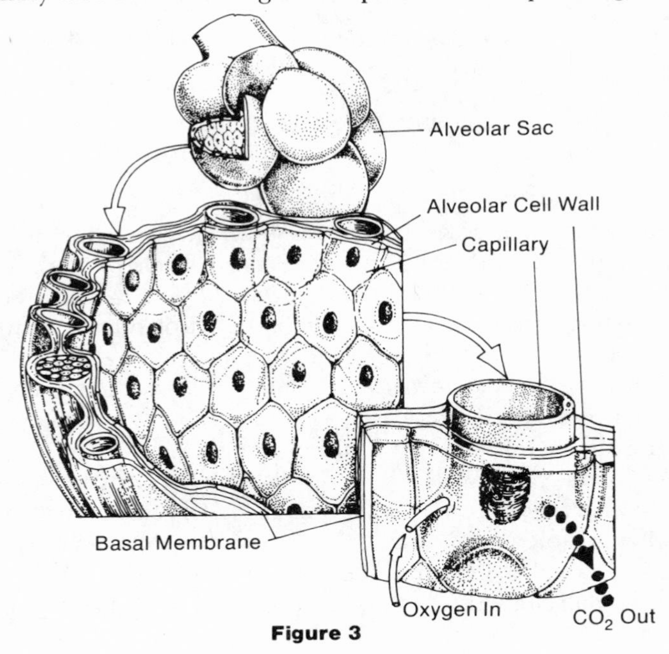

Figure 3

solubility. Incoming air is, of course, primarily nitrogen (78 percent) and oxygen (21 percent), plus fractional amounts of numerous other natural and man-made gases, including a tiny fraction (less than 1 percent) of carbon dioxide. As it comes into contact with the moist linings of the alveoli, some of its oxygen dissolves and begins the process of crossing over a three-part barrier composed of the cell wall of the alveolus, a specific kind of membrane known as a basal membrane, and the cell wall of one of the capillaries that lace the lungs.

Even after oxygen has crossed this tripartite barrier and entered the bloodstream, it must cross yet another membrane, the wall of the red corpuscle, in order to combine with the component of the blood, called hemoglobin, that is the specific oxygen carrier. As we discuss in later chapters, the toxicity of certain gases, such as carbon monoxide, stems from the fact that they, too, can readily cross over these same barriers to join with hemoglobin in a way that blocks oxygen from doing so, thus posing the risk of asphyxiation to anyone exposed to them.

Hemoglobin is transported by the bloodstream to different parts of the body, where it again passes through membranous walls to enter the specialized cells that make up the various organs. Inside these cells, hemoglobin combines with carbon, hydrogen, and other elements in biochemical reactions that release the energy needed to fuel growth, motion, and every other bodily function. The heat, water, carbon dioxide, and other wastes created by these reactions are then passed back out of the cell into the bloodstream or lymph system and eventually eliminated from the body in any one of a variety of ways. Waste carbon dioxide, for example, is carried back to the lungs in the bloodstream, diffused back across the capillary wall, basal membrane, and alveolar cell wall, and exhaled. Exhalations contain about 6 percent less oxygen and 5 to 6 percent more carbon dioxide than inhaled air.

THE GASTROINTESTINAL TRACT

Like the respiratory tract, the body's second major environmental contact point, the GI tract, is also equipped with numerous self-defense mechanisms, some quite similar to those of the respiratory system. First, of course, there are the basic sensations of smell and taste. (Taste is actually more a matter of smell than anything else.) Food that smells strange is less likely to be put in the mouth in the

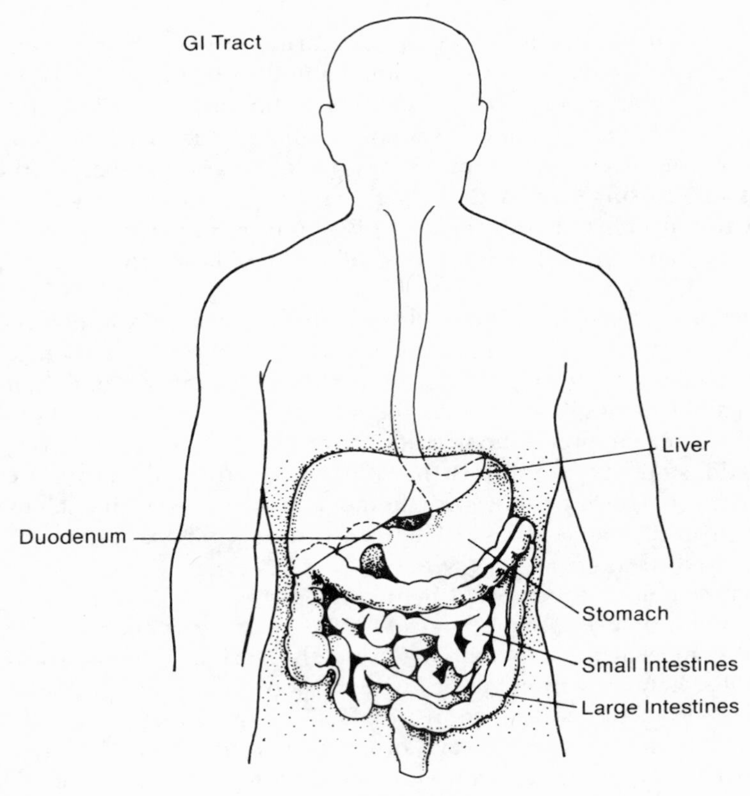

Figure 4

first place, and if it also tastes bad once it is in the mouth, it is less likely to be swallowed.

With the taking of food or drink into the mouth, a digestive process begins that lasts for some six to eight hours, and has as its main objective the liquefaction of solids and the breakdown of complex carbohydrates, proteins, and fats into the simple amino acids, fatty acids, and sugars that are able to cross over the intestinal walls and become absorbed into the bloodstream. Unlike the respiratory tract, in which waste gases are eliminated from the body through the same passageways they enter by, ingestion is a one-way process: raw materials are taken in at one end of the alimentary canal, and wastes are eliminated at the other.

A small amount of digestion takes place in the mouth, as digestive enzymes mixed with saliva begin to combine with food as it is chewed, but the process does not begin in earnest until food reaches the small intestine. Some breakdown of proteins takes place in the stomach, but it is primarily a holding tank, in which solids are kept for some three to four hours until they have been thoroughly liquefied. Hydrochloric acid secreted by some 35 million glands in the stomach lining contributes to this liquefaction process and sterilizes the stomach's contents by killing most of the bacteria that it contains. Other toxins susceptible to exposure to acids may also be effectively destroyed at this stage of digestion.

Like portions of the respiratory tract the stomach is also lined with a mucous membrane. Here it plays a crucial role by emitting a mildly antacid fluid that protects the stomach wall against its own highly acid secretions. As a result of this built-in protective device the wall of the stomach is far less vulnerable to chemical damage than virtually any other organ in the body. However, this antacid mechanism is not foolproof: when the stomach's acid glands start working overtime, often as a reaction to stress, the increased acidity that results can lead to the development of ulcers.

Once ingested food is completely liquefied, a thick valvelike muscle separating the stomach and the small intestine opens and the fluid mixture passes into the next segment of the alimentary canal. It is here, in the small intestine, that by far the greatest part of digestion takes place and ingested food completes its transformation into simple absorbable sugars and acids. To increase the amount of surface area available for absorption, the intestine is equipped with a wavy interior surface, lined with millions of small tuftlike projections, called "villi," somewhat analogous to alveoli.

The entire GI tract, especially the stomach and intestine, possesses remarkable powers of regeneration. The cells in the lining of these organs reproduce so rapidly that the entire lining is renewed every forty-eight hours. For this reason medical care for individuals with conditions like bleeding gastritis (a general inflammation of the stomach) focuses primarily on nothing more complex than assuring that the patient makes it through the first forty-eight hours without bleeding to death, because by then he or she will have grown a whole new stomach lining.

Unlike the respiratory tract, which has played all of its defensive cards once a gas has crossed over the wall of the alveoli into the bloodstream, the body still holds an ace in reserve after nutrients have entered the capillaries surrounding the intestine. Immediately after leaving the area around the intestine this blood is collected in a special vein, known as the portal ("carrying") vein. The portal vein then carries it directly to the body's single most powerful detoxifica-

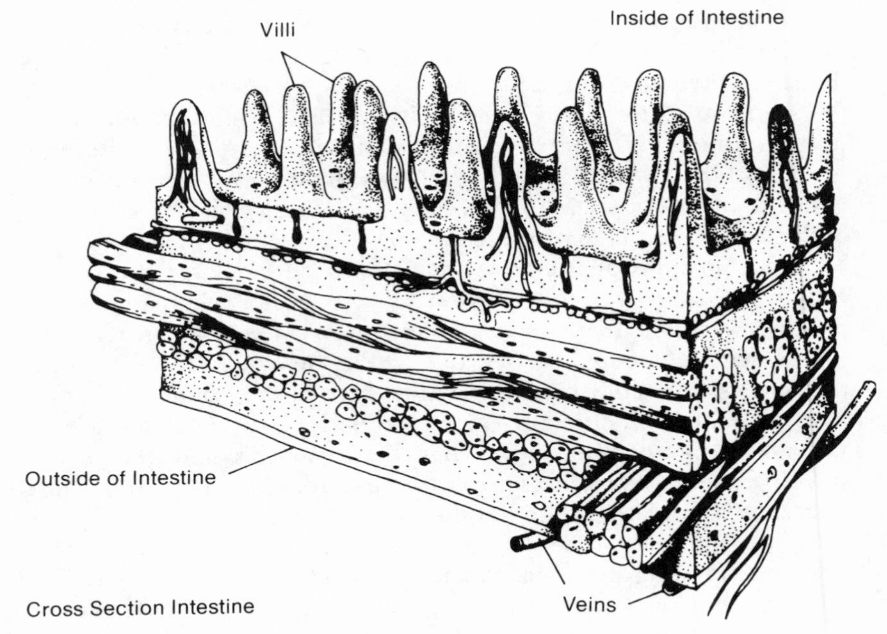

Figure 5

tion organ, the liver, which contains special cells whose sole purpose is to filter impurities out of the blood.

The liver is a large organ located just under the lower ribs, next to the stomach. It serves two dozen clearly distinguishable functions, ranging from manufacturing bile and cholesterol to storing a form of blood sugar called glycogen to dispatching bile to the gall bladder for use in digestion. But from an environmental health standpoint its most important purpose is detoxifying the nutrient-rich blood flowing away from the GI tract and other parts of the body, including the lungs and the skin. The liver typically performs this function by transforming the chemical structure of a hazardous compound, usually by adding a glucaronic or sulfuronic acid molecule to it to make it soluble enough to be passed through the kidneys and out of the body in the urine.

Nature's propensity for designing with safety factors has been repeated to a small degree in the liver. An otherwise healthy person can function with a small degree of liver impairment. Furthermore, although chronic exposure to toxins can cause an irreversible condition known as cirrhosis, in which healthy liver tissue is replaced by fat and scar tissue, the liver has good regenerative powers, and can recover from most kinds of limited damage caused by temporary exposures to a wide variety of hazards.

The very last stage of digestion, which takes place in the large intestine, is recovery of most of the water added in earlier stages. Once the water has been reabsorbed, all that is left are waste products, the feces, and when these are eliminated from the body the digestive process is complete. The entire cycle takes about twenty-four hours on the average.

THE SKIN

Since the skin is so much more exposed to the ambient environment than any other organ, it's hardly surprising to find that skin disorders are among the most common responses of the body to environmental irritants and hazards. In the occupational environment, where concentrations of toxic compounds are generally higher than in the environment at large, some 60 to 90 percent of all medical complaints involve the skin.

Fortunately, the skin's exposure also means that disorders of this organ tend to be easy to detect and diagnose. For this reason, and because the skin is also a highly resilient organ, skin diseases are rarely fatal. (The main exception is a type of cancer called malignant melanoma.) In fact, in many cases skin rashes and other irritations are regarded more as useful warning signs of environmental hazards than as illnesses.

The skin has three main functions: to contain body fluids, to eliminate heat generated in the body by the never-ending biochemical reactions that take place inside it, and to protect against environmental threats of all kinds—thermal, electromagnetic, physical, and chemical.

Since the interior of the body is mostly water—even our bones are 20 to 25 percent water and most other organs have a far higher liquid content—there would be no way to maintain life if our body fluids weren't contained within an impermeable barrier that blocked them off from the dry environment of land and air that evolution has brought us to. In cases where much of this barrier is accidentally destroyed, as in severe burns, this is precisely what happens: the body's fluids drain away so rapidly that lethal chemical imbalances in the body occur, the vascular system collapses, and life ceases.

To maintain a constant body temperature, the skin functions as an effective cooling system by shedding heat through simple radiation and by evaporation of sweat. In radiative cooling, heat generated by metabolic processes is passed close to the skin surface by the blood. From there it radiates out into the external environment. This simple mechanism is complemented by the evaporative cooling system that we know as perspiration. When we perspire, a thin film of moisture is exuded onto the skin's surface by sweat glands buried just below it. The chemical reaction that occurs as this liquid becomes a gas—water vapor—absorbs heat, drawing it away from the body. If the air around the body is already so humid that more water vapor cannot enter it, this process is slowed or halted altogether, and we quickly begin to feel so uncomfortable that we automatically reduce our physical activity or turn on a fan to circulate less-saturated air around us to carry water vapor and heat away with it.

Without these natural, highly efficient cooling systems, as the thermometer outside the body began to climb, so would its internal temperature. After a fairly short interval, a day or two at most, of internal temperatures in excess of 38 to 43°C, the very structure of the body's proteins would begin to change, and life would no longer be possible.

What we casually refer to as skin is actually a series of different layers of tissue, organs, and glands. The innermost layer, known as

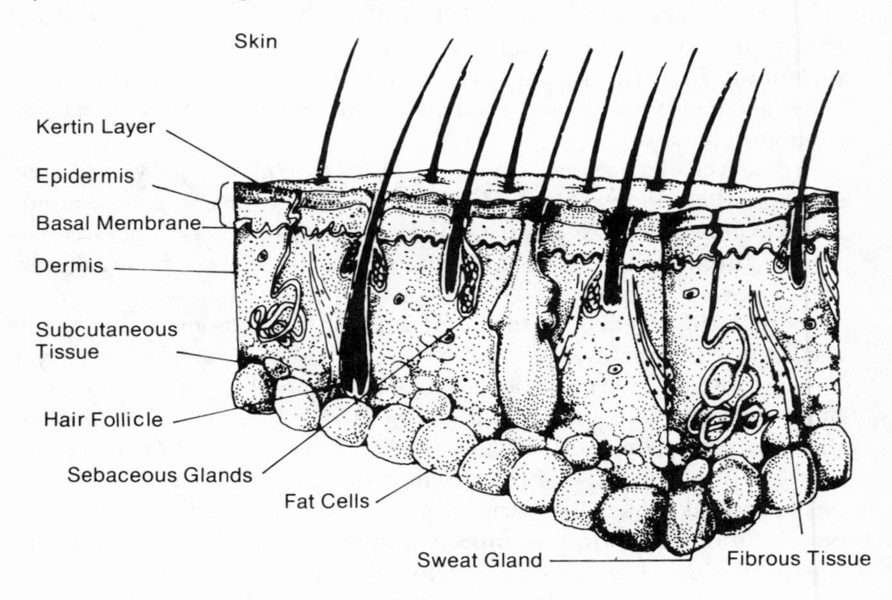

Figure 6

subcutaneous tissue, bonds the layers above it to the body. Skin diseases or injuries that penetrate to this layer pose the greatest threat to life by permitting the escape of body fluids and by exposing delicate interior body tissues to the risk of infection by airborne organisms. The layer directly above the subcutaneous layer is called the dermis. It contains all the glands, blood vessels, hair roots, and nerve endings that serve the skin as a whole. Above the dermis is the basal layer, which contains specialized cells that perform two functions unique to the skin's role as our most exposed organ. One type of cell in this layer, known as a melanocyte, helps protect us from exposure to hazardous amounts of ultraviolet light by producing a natural sunscreen pigment, known as melanin, which colors skin with hues ranging from light reddish-pink to deep black. As well as providing the basic skin hue we are born with, melanocytes can respond to increased exposure to sunlight by producing more pigment, a process we call getting a tan. However, at the same time that it provides pigment to prevent us from overexposure to ultraviolet light, the body must be careful not to block out sunlight totally, since we need to absorb a small amount of it to form vitamin D, required for healthy bone development.

The second major function of the basal layer of the skin is to provide the cells that form the outermost layer of skin. As it grows, a specific type of basal cell detaches itself from the basal membrane and slowly migrates outward, toward the exposed exterior surface of the skin. However, because the layer of skin just above the basal membrane, called the epidermis, contains no blood vessels, it cannot nourish basal cells after they leave the basal membrane. So, as they migrate outward they die. At the same time, they become flatter and flatter as they move, until, by the time they reach the outermost layer of skin, known as the keratin layer, they are completely flattened into a shape ("squamous," in medical parlance) that resembles small paving stones.

Because of this constant migration of basal cells to the outermost layer of the skin we are completely covered by a thin layer of dead cells that provides an excellent lightweight buffer zone against many routine environmental insults. In areas that get a lot of use, like the palms of the hand and the soles of the feet, the body responds to the increased wear and tear on the skin by thickening the keratin layer, causing calluses. Moreover, the keratin layer constantly regenerates itself, at the rate of a complete replacement every month or so. This means we can tolerate a good deal of minor skin abrasion, irritation, and other damage without suffering significant harm, and

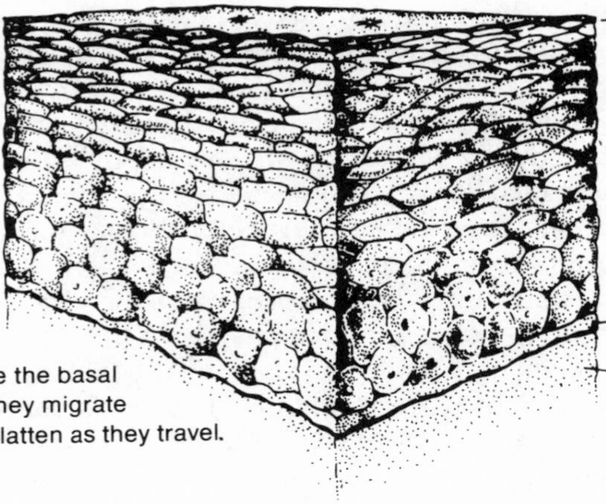

Figure 7

in many cases without even noticing it. On the whole, our skin is fairly impervious to environmental hazards. Oil produced by glands in the dermis known as sebaceous glands help make it resistant to water and mild acid solutions. It is a fairly good barrier against most gases. (If it weren't, we wouldn't need lungs, since we would be able to absorb oxygen directly through the skin.) Most bacteria and viruses, and a good three-quarters of the chemical substances we encounter in the biosphere, are successfully repelled by intact skin.

However, it is not absolutely impermeable. Strongly acidic or alkaline solutions and most solvents and detergents have the potential to enter the body through the skin, especially through the natural openings provided by hair follicles and pores. The skin on certain areas of the body, especially the face, scalp, stomach, and genitals, is quite a bit more absorptive than areas like the palms and soles.

And there are the inevitable trade-offs; when we work or exercise hard, our natural cooling mechanisms tend to make the skin hotter and moister, since the capillaries just below its surface dilate so that blood can travel through them more rapidly, and we begin to sweat more. Under these conditions absorption of foreign substances from the ambient environment is maximized. Unfortunately, these are precisely the kinds of skin conditions that heavy manual labor in a plant or factory is likely to cause, and it is in the occupational environment that exposure to high concentrations of many industrial toxins is most commonly encountered. Skin that has been scraped or cut is thousands of times more absorptive than intact skin, for self-evident reasons. Under the most adverse circumstances, especially when dealing with highly concentrated solutions of solvents and other toxins that easily penetrate the skin, skin absorption can be just as hazardous as any other type of exposure:

> A tank truck driver was preparing a cotton defoliant by mixing diesel oil with pentachlorophenol. As he was drawing the concentrated chemical out of a drum, the spigot accidentally toppled back. He reached in with his bare hand to regain the spigot. Although he washed immediately, he became acutely ill and died the next day.[1]

All in all, the skin is an ingenious solution to many of the problems posed by the interaction of a relatively sensitive organism like the human body with an often hostile environment. We may have sacrificed some of the protection afforded by more cumbersome alternatives, like the armor plating of the turtle or the dense blubber of the whale, but we've also gained an advantage in mobility and adaptability to widely varying environments in return.

CANCER AND THE ENVIRONMENT

The statistics about this most dreaded of all diseases are certainly sobering enough: one in every four Americans alive today will develop some form of cancer in his or her lifetime. Cancer is currently our second leading cause of death. Heart disease is first, accounting for nearly 40 percent of all annual fatalities in this country. In 1981 some 665,000 new cases of cancer were diagnosed, and approximately 420,000 Americans died of cancer—48 every hour, around the clock, every day of the year. And in many other ways cancer is certainly worth fearing. Depending upon the type of cancer and the organs involved, a death by cancer can be one of the most debilitating, painful, and disfiguring a human being can suffer.

Because cancer can be such a terrifying affliction, and because the emotions it arouses can be such powerful and irrational ones, misconceptions abound about cancer and the role that the industrialized environment plays in causing it. First, cancer is not a new disease. It antedates the Industrial Revolution by well over two thousand years, at least. The very word "cancer" (meaning crab) was first used to describe the disease* by the Greek physician Galen in the second century A.D. because the swollen veins radiating away from a typical cancer tumor reminded him of the appearance of a crab's legs.

Nor is cancer strictly a product of man-made substances. There are a number of natural carcinogens in the environment, some of which are much more common and potentially much more hazardous than most man-made carcinogens. For example, in humid climates a mold called *Aspergillus flavus* that grows on groundnuts (e.g., peanuts) and on some grains (e.g., rice, corn) can cause a virulent type of carcinogen known as aflatoxin. In parts of Africa and Southeast Asia, numerous studies have suggested a strong correlation between aflatoxin contamination of the diet and high levels of liver cancer.

Another common misunderstanding about cancer is the idea that it consists of cells that "grow faster" than other cells. Much of what happens when cells become cancerous, especially at the molecular level, is still very much a mystery to us, but we do know that cancer cells do not grow any faster than other, perfectly normal, cells of the rapidly dividing type. Instead, it is more accurate to say that cancer cells grow in an *uncontrolled* fashion. To be specific, they appear to lack a property shared by all healthy cells known as "contact inhibition." Contact inhibition means that a cell will grow only until it comes into contact with another cell; then it stops. Cancer cells, lacking this intrinsic trait, just keep on growing, dividing, and continuing to grow, a process that leads to the swellings we call tumors. In doing so, they hog nutrients, robbing other cells of the substances they need to survive. They also appear to emit special compounds that actively interfere with growth and metabolic processes in normal cells. As a result of this combined loss of nutrients and attack by growth-inhibiting substances, healthy cells in the vicinity of cancer cells tend to stop growing and eventually die.

Still another popular misperception about cancer, closely related to those about its growth processes, is that cancer is always the result of carcinogens that enter the body from the external environment. In some cases, such as with cigarette smoke, this is true. But in other instances, instead of being the passive victim of assault by carcinogens attacking from outside it, the body itself is the source or at least a significant accomplice in the onset of cancer. First of all, it can and does spontaneously create tumors on occasion, without any prompt-

*Actually, cancer is not one single disease, but well over a hundred distinct illnesses, all sharing certain common characteristics.

ing from environmental factors, perhaps as a result of genetic traits. In some cases these tumors turn out to be malignant. Furthermore, in other instances substances in the ambient environment that are associated with tumor induction do not actually become carcinogenic until they have been metabolized in the body, usually by certain liver enzymes. Benzene is a good example. In these cases any cancers that result are caused by an unfortunate collaboration between substances in the environment and those in the body.

Possibly the most common cancer misconception is that most industrial chemicals and byproducts are carcinogenic. In fact, of the seven to eight thousand substances that have to date been adequately tested for their carcinogenic potential, only about 10 percent have been found to be carcinogenic in lab animals. Considering the fact that careful testing for carcinogenicity is a relatively recent process, and that many substances selected for testing were already suspected to be carcinogenic for other reasons, this is a surprisingly low percentage. Using this figure of 10 percent to predict the number of the sixty-five thousand chemicals in common use in this country that may turn out to be carcinogenic, we get a total of some sixty-five hundred. This is certainly no insignificant amount, but it is a far cry from the total of thirty-five thousand or so chemicals that the EPA has labeled toxic, and it is clearly in no way supportive of the view that "everything" is carcinogenic.

One reason why it often seems as if every chemical we use in this society is carcinogenic is that the process by which substances are selected for carcinogenicity testing is hardly a random one. In most cases, compounds picked for this testing have already given us indications—from occupational exposures, from structural similarities to known carcinogens, or for other reasons—that they may be cancer causers. This is why the committees that pick substances to be tested have about a 50 percent success rate so far in selecting those that do turn out to be carcinogenic. But the fact that well-informed researchers forearmed with good clues about a compound's potential carcinogenicity turn out to be right half the time hardly means that half the chemicals in the environment are causes of cancer.

It is difficult to say how many of these sixty-five hundred projected animal carcinogens will also turn out to be human carcinogens. Since we cannot test suspected carcinogens directly in human subjects, we will have to wait while sufficient clinical and epidemiological data—most of which will come from occupational environments—accumulate to support or refute the bioassay data, and this process will invariably be dogged with controversy. To date, only about twenty-eight substances have been classified as *known* human carinogens. Another sixty or so are strongly suspected to be human carcinogens. The twenty-eight known carcinogens are:

Aflatoxins
Aminobiphenyl
Arsenic compounds
Asbestos milling, installation, and tear-out
Auramine manufacturing
Benzene
Benzidine
Bis (chloromethyl) ether
Cadmium oxide
Chlorambucil
Chloromethyl methyl ether
Chromate-related industries
Cyclophosphamide
Diethylstilbestrol
Hematite mining
Byproducts of the manufacture of isopropyl alcohols
Melphalan
Mustard gas
2-Naphthylamine
Nickel refining
N, N-Bis (2-chloroethyl)-2-Naphthylamine
Oxymetholone
Phenacetin
Certain soots, tars, oils, pitches, and smokes
Radioactivity
Thiotepa
Ultraviolet light
Vinyl chloride[2]

However, the near unanimous consensus of cancer experts is that substances that are carcinogenic in test animals usually hold some potential for causing cancer in humans, too. Humans may be more sensitive to the compound than animals, or less, cancers may develop in different organs in humans than in test animals, or at different rates, but the bottom line is that extrapolating from other species to man has been shown to be a reliable practice over and over again. So we can eventually therefore expect to find that most of that projected sixty-five hundred animal carcinogens will also turn out to pose some carcinogenic risk for humans. But since cancer has such a long latency period, up to forty years, we won't know for sure until well into the next century just what the exact percentage is. In addition, there's the sobering fact that we're creating upwards of one thousand new chemicals annually, and we're only able adequately to test less than one-third that many each year for carcinogenicity. Without the highly educated guesses of the scientific screening panels that decide which substances need to be tested right away and which can wait, we'd be stuck with a dangerous knowledge gap we could never hope to close.

The most erroneous cancer misconceptions of all are those bred by the often-heard statements to the effect that "80 percent of all cancers are due to environmental causes." Without further explanation of what the term "environment" means, such an assertion, coupled with the many other common misconceptions about cancer that we've already touched on, leads people to assume mistakenly that not only is cancer strictly a product of industrial development, that virtually every chemical man devises and releases into the environment is potentially carcinogenic, but also that as a result of this development members of the industrialized societies of the world live in a boundless sea of carcinogenic perils that are ubiquitous in their air, water, and food.

This isn't the case. First of all, the term "environment" as used in the assertion above does not refer only to the ambient air, water, and soil of the biosphere. It is a much broader term, so inclusive as to be in many cases almost meaningless. At its broadest, the term "environment" used in this way simply denotes any tumor incidence above that which can be expected to occur naturally in any given population. In other words, as we explained in Chapter 1, in any group of people there will always be some degree of cancer that occurs spontaneously, regardless of what conditions are like in the ambient environment. Scientists calculate this level by comparing cancer morbidity and mortality statistics for a wide spectrum of population groups living in varying types of cultures in different geographical areas. In this way they can determine an incidence level that can be expected to occur for any type of cancer in any population, no matter what personal, cultural, and industrial conditions are like. This is called a "base-line" level, and any incidence of cancer above it is attributed to "environmental" causes.

Clearly, used in this way the term "environment" is so inclusive as to mean basically "anything other than normal, natural, inevitable physiological processes," or "everything external to the body," including personal habits and life-style, cultural and sociological patterns, medications, occupational settings, as well as the urban environment and the natural biosphere. Of all these factors, personal habits and cultural patterns are far more important determinants of cancer incidence than industrial pollutants at large in the biosphere in the developed nations.

For example, cigarette smoking—which must be defined as an "environmental" carcinogen under this all-inclusive definition—is easily the single greatest contributor to the incidence of lung cancer in this country today. In 1900 the overall mortality rate for all cancers in the United States was about $^{64}/_{100,000}$. Today it is well over twice as high—about $^{150}/_{100,000}$—and most of that increase can be attributed to lung cancer fatalities caused by smoking cigarettes. Experts estimate that, in men, smoking causes approximately 40 percent of all cancer cases; among American males, lung cancer is the leading cause of all cancer mortalities, accounting for some sixty-five thousand deaths annually. Among American women, lung cancer is currently the third leading cause of cancer deaths, causing about twenty thousand or so mortalities a year, but because more and more young women are taking up smoking, it is rapidly rising toward first place, and by 1983 will in all likelihood have surpassed the current first-place holder, breast cancer (now responsible for about thirty-three thousand deaths annually). Cigarette smoking also appears to be a significant factor in the development of cancer of the mouth, throat, larynx, and bladder.

Ambient industrial pollution in the atmosphere, particularly in large metropolitan areas, certainly hasn't helped reduce the increase in lung cancer, and is instead much more likely to have played a significant role as a "cocarcinogen" in promoting the rising incidence of lung cancer. But it is clearly not the primary cause of this increase.

After smoking, the next most significant factors contributing to cancer incidence in this country are alcohol consumption and other complex, poorly understood dietary patterns, such as ingestion of high levels of carbohydrates and fats, low levels of fiber and fresh fruits and vegetables, and high levels of nitrates. Alcohol consumption has been linked to increases in the incidence of cancer of the throat and larynx. High-fat, low-fiber diets appear to be related to increased incidences of cancer of the large bowel,* possibly, accord-

*Colon cancer and rectal cancer are usually listed separately in tabulations of cancer mortality statistics. Some cancer experts, however, argue that they should be combined into one category, "large bowel cancer," which would then automatically being to some experts, because such diets tend to be associated with longer retention of feces in the bowel and thus greater exposure to whatever carcinogens they may contain.

After smoking and diet, the next most significant carcinogenic risk factor in a modern industrialized society is probably occupational exposure in the workplace, which accounts for somewhere between 1 percent and 5 percent of all cancers. Hereditary influences and extended exposures to sunlight and other kinds of natural radiation are of a similar level of significance, each causing an estimated 5 percent of all cancer cases.

Although percentages can be misleading, since all of these variables and many others overlap and interact with one another considerably, the three main cancer risk factors, smoking, diet, and occupation, probably account for at least half and perhaps as much as three-quarters of all cancer cases. Clearly, if the term "environment" is defined so as to exclude personal habits like smoking and drinking, cultural patterns like diet, and exposures to carcinogens that are restricted to the workplace, then the amount of cancer that can be attributed to truly "environmental" pollutants is far less than the touted 80 percent.

In short, as Dr. John Higginson, former head of the World Health Organization's highly respected International Agency for Research on Cancer, has put it:

> Available epidemiological data indicate that for man the most widespread and important carcinogenic stimulus so far recognized is cigarette smoking which affects the mouth, lung, larynx, oesophagus, and possibly bladder and pancreas. In the lung, the effects of cigarette smoking are potentiated by air pollution, asbestos, and radon. In addition, certain alcoholic beverages may cause cancers of the mouth, oesophagus, liver and possibly rectum and may also potentiate the carcinogenic action of cigarettes. The role of diet in human cancers is less clearly defined, but obesity, over-eating, and dietary patterns are apparently associated with increased frequencies of certain cancers, including colon and possibly breast. . . . Thus the evidence would suggest that a male who lives in the country, does not smoke, eats and drinks with moderation, and reduces his exposure to sunlight may reduce his risk of cancer by at least 30–40 percent or pos-

---

come our leading type of cancer, with some one hundred thousand new cases each year, about half of which are fatal.

> sibly more according to the country in which he lives. The corresponding figures for females are somewhat less. Nonetheless, such massive changes in lifestyle must be started early in life, as the later these measures are taken the less they have to offer.[3]

As Dr. Higginson also points out in the same article, additional empirical support for his identification of smoking, alcohol, and diet as major contributing factors in the development of a majority of cancers comes from the fact that nonsmoking, teetotaling groups like Mormons and Seventh-Day Adventists (who are also vegetarians) exhibit cancer incidences that are significantly lower than national averages, as much as 50 percent lower in the case of the latter group.

Still more evidence for the preeminence of personal habits and cultural patterns over ambient industrial pollutants as primary carcinogens shows up when cancer rates in different but equally industrialized nations are compared. If industrial development were the primary determinant of cancer incidence in the world today, one would logically expect to find that countries of similar technological advancement have similar incidence and mortality rates for the same kinds of cancers. But this is most definitely not the case. In the United States and Japan, which both have roughly equal levels of industrialization, pollution problems, and record-keeping systems for recording vital statistics, as well as a high proportion of male smokers, lung cancer rates differ sharply: $^{44}/_{100,000}$ for the United States versus slightly less than $^{16}/_{100,000}$ for Japan. The situation is reversed for stomach cancer. In the United States the stomach cancer rate has been dropping steadily since the early 1930s and now stands at just under $^{15}/_{100,000}$. Japan's rate of stomach cancer is an astonishingly high $^{95}/_{100,000}$. There are many similar examples of disparities between more or less equally industrialized nations. Scotland's lung cancer rate is quite high; Portugal's is quite low. Among nonwhite American males the incidence of pancreatic cancer is high; among Italian males it is quite low. Breast cancer rates are high in Holland, low in Japan. Moreover, as numerous studies have demonstrated, when individuals leave one country and migrate to a different one they leave their old cancer risks behind and assume those of their new home: Second- and third-generation Japanese-Americans, for example, lose their tendency to develop stomach cancer and acquire an elevated risk, like other Americans, of lung, prostate, colon/rectal, and bladder cancer instead. Obviously, other factors besides the degree of industrialization of an area, factors like personal habits, cultural patterns, type of industrialization, and perhaps genetic variables, are also at work.

CANCER TREATMENT

At one time or another virtually every treatment imaginable has been employed in the fight against cancer. Special diets, salves, electrical devices, exercise regimes, megavitamin injections, bloodletting, acupuncture—all these and thousands of other approaches have been employed, generally with little or no success. As we see today in the popularity of laetrile, a substance derived from apricot pits that has no scientific efficacy against any form of cancer, mankind's desperate search for a magical remedy against this frightful disease continues unabated.

Today, the effective medical response to cancer depends first of all on early diagnosis. Because cancer is a disease that spreads throughout the body ("metastasizes") in most cases, becoming a greater peril with each new organ system it invades, early detection and treatment is the foundation of any successful cancer protection program. Pap smears, chest X rays, breast exams, proctological exams, and a myriad of other diagnostic techniques constitute the bulwark of the fight against cancer.

Once detected, cancer treatment involves a triad of basic approaches—surgery, radiation, and chemotherapy—used independently or in conjunction with one another. Generally, surgery is used when the tumor is in a specific and accessible location, before the cancer has spread to other parts of the body. Breast cancer is a typical kind of cancer where surgery is a common treatment response, though the extent of the surgery performed in many cases of breast cancer has been significantly reduced in the last few decades, and radiation and chemotherapy are much more widely used as adjuncts to breast cancer surgery than they were not long ago.

Radiation therapy was being used to treat cancer within weeks after the discovery of X rays in the late nineteenth century, and with the post-World War II development of high-energy X-ray beams that could be very narrowly and accurately focused on the tumor site, it became an even more effective treatment mode. Radiation therapy with X rays or with radioactive substances that are ingested or injected into the body tends to be used most commonly with cancers that attack organ systems not easily accessible to the scalpel, like the lymph system, or those that cannot be excised because nature hasn't provided much of a safety factor in the form of excess organ capacity.

Chemotherapy—that is, the use of any drugs or chemicals—is both the oldest and newest member of the cancer treatment triad. The contemporary chemical armamentarium was greatly expanded just after World War II by derivatives of nerve gas research, and in succeeding decades by other, equally powerful chemical agents, most of which are so toxic they pose the risk of hazardous side effects of their own in cancer patients if the treatment process is not managed with great care. In general, chemotherapy is most appropriate as a treatment for systemic cancers, such as leukemia, that are not restricted to one part of the body.

Chemotherapy is currently the area of greatest promise in cancer treatment, and most of the most encouraging advances achieved over the last few decades, such as the great improvement in childhood acute leukemia survival rates, are due to improved chemotherapy techniques.

We seem to be doing something right with our approach to cancer treatment, since survival rates for the most common cancers are slowly improving. Ten years ago the percentage of patients who lived at least five years after major cancers was 9 percent for lung cancer, 65 percent for breast cancer, 46 percent for colon cancer, 42 percent for rectal cancer, 2 percent for pancreatic cancer, 57 percent for prostate cancer, and 62 percent for bladder cancer.

Today, according to the American College of Surgeons, those rates have all improved, by an average of almost 6 percent. The five-year survival rate for lung cancer is now up to 11 percent, for breast cancer it is 73 percent, for colon cancer it is 50 percent, for rectal cancer it is 49 percent, for pancreatic cancer it is 3 percent, for prostate cancer it is 68 percent, and for bladder cancer it is 70 percent. Improvements in the five-year survival rate for some less common cancers is even greater. For example, acute leukemia survival rates have risen to 18 percent from 3 percent a decade ago.

3

OCCUPATIONAL HARBINGERS

Though the environment at large is increasingly permeated with man-made substances, many toxic, it is often in the workplaces where these substances are made, used, and stored that the highest exposures to them occur. All too frequently, despite the development in recent years of numerous systems intended to prevent this from happening, the way we first learn of a compound's potential toxicity is only after it has already caused disease in those who work with it. The men and women involved in such cases are truly society's canaries: like the caged birds once carried into mines to warn by their death of the presence of noxious gases, they are occupational harbingers for the rest of us.

The interdependence of the environment, combined with the fact that industries inevitably release some amount of the substances they use into the biosphere, no matter what kind of emission controls they employ, means that the discovery of a hazard in the workplace automatically raises the question of hazards to the public at large. Typically, this is a difficult question to try to answer, for a number of reasons. First, determining the level of exposure that is toxic in the workplace—to say nothing of the ambient environment—can be quite complicated. Until very recently few companies kept detailed records of where each employee worked and what substances he or she came into contact with. Going back long after the fact to reconstruct how much of a dose of compound X a particular worker might have received is in these cases a matter of informed guesswork. Moreover, workers are typically exposed to complex, changing mixtures of a number of substances rather than just one; identifying the real culprit(s) becomes virtually impossible in such circumstances.

Second, though workplace concentrations of a substance are typically much higher than those found in the ambient environment, workers are in the presence of these substances for only 40 hours a week, on the average, while the general public is exposed to whatever contaminates the ambient environment all the time—168 hours a week.

Third, a workforce population is not a representative cross section of the population at large, in terms of overall health. Workers on the whole are healthier, since the general public includes many hypersensitive individuals—the very young, the very old, and those with preexisting disorders, such as lung diseases—that may make them particularly susceptible to toxic compounds. All other things being equal, workers can usually tolerate higher exposures than the public can.

Fourth, the very fact that the workplace is known to be hazardous has led to the development in many industries of special safety and health precautions for workers, such as the use of respirators, protective clothing, and the administration of regular, specialized health screening tests designed to detect toxic effects early, so they can be forestalled. As a result, though he or she may be physically closer, more of the time, to hazardous compounds, a worker may also in fact be much safer than a member of the general public who has no special protection at all and who may not even know what compounds he or she is exposed to or what the levels of this exposure might be.

Finally, we tend to think of toxicity as being directly related to dose: The more you get, the sicker you become. But, as we explained in Chapter 1, dose is a function of both the concentration of the substance in the environment and the duration of exposure to it. Sometimes a long exposure to a low concentration can be much more hazardous than a series of short exposures to a much higher concentration, particularly if the substance happens to be one that is retained in the body for long periods of time. Furthermore, some substances are so toxic that virtually any exposure is hazardous; they do not have to reach a certain "threshold" to be dangerous.

ASBESTOS

Asbestos is a natural mineral that is extremely resistant to heat. It is so fibrous that it can be woven into a tough nonflammable cloth when mixed with cotton. It is also light, strong, durable, absorbent, and cheap.

Asbestos has been known and used in a minor way since antiquity. Throughout history, chroniclers have marveled at the amazing fabric that could when soiled be tossed into the nearest fire and retrieved, spotless and undamaged, a few moments later. The word "asbestos," meaning unquenchable, probably comes from its use as a perennially unburned lamp wick.

Until the Industrial Revolution asbestos was little more than a semimagical curiosity. But as the age of the horse gave way to the age of the machine, the value of a plentiful substance that was light, tough, resistant to heat and corrosion, and inexpensive slowly but steadily opened up new industrial applications. A century ago worldwide production of asbestos was about five hundred tons per year. Today it exceeds 5 million tons. No other commodity, including oil, has seen such a rapid jump in production in the last one hundred years.

Most of the world's asbestos is mined in Quebec, South Africa, Finland, and the Soviet Union. The United States produces less than 4 percent, but manufactures 25 percent of the world's asbestos products, especially those used in construction.

Today, asbestos is used in thousands of materials, appliances, and other items, ranging from sewer pipes to corn poppers. A few of the more common applications include wallboard, roofing material, brake shoes, clutch parts, mailbags, aprons, movie screens, theater curtains, stove linings, and heaters. Because asbestos fibers are so absorbent, they have also been widely used since the turn of the century to filter a variety of foodstuffs, especially such beverages as beer, wine, and a number of different brands of soft drinks, as well as some types of drugs.

Working with asbestos in milling operations, where it is separated from the raw ore, cleaned, sorted, and packed, or in any of the dozens of different manufacturing processes where it is unpacked, poured, ground, and mixed was until very recently a notoriously dirty, dusty job. During these procedures the fibers break up into small particles that are easily carried about in the air and inhaled.

By the early years of this century concern was beginning to appear in England that inhalation of asbestos dust might be a health hazard. But the signs of asbestos-related diseases were for some time difficult to distinguish from tuberculosis, and it wasn't until 1924 that firm evidence that asbestos alone caused scarring of the lungs and impaired breathing was published in an English medical journal. By 1927 the term "asbestosis" had been coined,[1] and three years later Parliament was officially informed that inhalation of asbestos dust posed a significant threat to the health of industrial workers, one that could be avoided by better dust-control measures. By the mid-1930s the first indications were appearing in Great Britain and the United States that asbestos inhalation was also linked to lung cancer. By 1955 that link had been conclusively established. Not until the 1960s was the connection of asbestos with another, rarer form of cancer of the lining of the lungs and the abdomen, called mesothelioma, also confirmed.

By the late 1930s government health officials in the United States were actively working to abate the problem of asbestos exposure in the workplace. A 1938 study of the asbestos textile industry by the U.S. Public Health Service[2] concludes that asbestos concentrations should be kept below 5 million particles per cubic foot (at least ten times the current exposure limit), and advises the use of vacuum hoods to help maintain that level. Had this trend of growing awareness of the asbestos problem continued, it seems likely that appropriate controls would have been brought into widespread use throughout the 1940s and 1950s, especially as it became clearer that asbestos also caused lung cancer.

However, with the outbreak of World War II, the problems of what seemed at the time like a few thousand industrial workers in the developed nations quickly slipped into obscurity, and government recommendations for controlling the dust problem were quickly forgotten, especially in the shipyards, where they would have slowed production considerably. Thus the stage was set for one of the greatest occupational health disasters of the century.

Because the war was fought in great part on the sea, a tremendous surge in demand for ship construction and repair work occurred, and shipyards on both coasts boomed. In December 1943, at the height of the war, a ship a week was coming off the ways in many yards, and nearly 2 million Americans were employed in them. By the war's end a total of nearly 4.5 million men and women had been employed in shipyards across the country.

There is no greater danger to a ship than a fire at sea, and to help combat it, naval and merchant marine ships are heavily insulated with asbestos. In U.S. shipyards blocks of preformed asbestos were cut to fit boilers, doors, machinery, and walls, then wired into place, or loose asbestos fibers were stuffed into cloth jackets that were sewn around pipes or conduits. Seams and joints were finished with a cement containing more asbestos.

The work of cutting, crushing, and sanding asbestos filled the air with respirable fibers. Only a few thousand shipyard workers in World War II were insulators, but the number exposed to dangerous levels of asbestos was far higher. To expedite the war effort government recommendations on allowable fiber concentrations in the air were in essence abandoned.

After the war, shipyard activity dwindled rapidly. Within a year only a few hundred thousand workers remained. However, many of the others found work in the postwar construction boom, and this meant that in many cases they were once again working with asbestos products, and were once again regularly inhaling high levels of fibers as they cut or sanded wall panels, floor tiles, asbestos-cement pipe, drywall "mud," plaster, stucco, and above all, insulation, all commonly made with asbestos.

George Fuqua is one of those workers. He began work as an insulation helper when he was twenty-two, in 1940. During the war years he worked in nearly every shipyard in the San Francisco Bay Area—Kaiser, United Engineers, Bethlehem Steel, Hunter's Point, Triple A, Matson—installing asbestos insulation. His wage was sixty cents an hour. After the war he joined Local 16 of the International Association of Heat and Frost Insulators and Asbestos Workers, and put in another twenty-five years of labor until, late in 1969, he began to notice something was wrong.

> I was the oldest man in the shop. It got to where my wind was going all the time, and I just couldn't get it back. I couldn't figure it out. Am I just getting a little older? Is it just me? Am I just slowing down? Then, on December seventeenth, just before Christmas, they called me in and laid me off. I got shot out of the saddle just when I was finally making it, after all those years. They knew I had asbestosis. Now I'm wiped out.
>
> It's a disgrace. All I've done is work hard like I was supposed to. During the war I was deferred because of my job, but now it looks like I would have been better off going overseas and getting shot at than working with asbestos. These shipyard insulator cases are just the tip of the iceberg. All the other crafts ate just as much of the stuff as the people like me who were

applying it. Welders, shipwrights, painters—every craft right down the line. You see, you're in compartments in the hold of a ship, and back then they had canvas ducts, with big fans, up on deck, and they would blow air down into the compartments where we were working. They had to do that because the welders would be down there burning, so they had to run air down there every time they had a welder working. And when we cut the asbestos, it didn't just sit on the floor. It got picked up in those ducts and it circulated. It blew everywhere—up, down, out, everywhere. Sometimes I remember it was so thick you couldn't see the bulkhead ten feet away. It was just a big dust storm down there.

We wondered if it was good for you, all that dust. I remember once asking a guy from Johns-Manville* if it was OK. I said, "Are you sure this stuff won't hurt you?" He said, "You could eat it."

If you started to complain that your lungs were hurting, or that you couldn't breathe right, they'd weed you out, the companies, tie a can to your tail. I saw it happen. I was supervising a job and the man that owned the company told me, "If so-and-so gives you any trouble, if he gripes about anything at all, send him to the office and we'll straighten him out." That was how they scared the workers into going along with it and not making a fuss. Now, today, that particular worker that the owner was talking about, he's dead from asbestosis. The only thing I do these days is go to funerals of friends who've died from working with asbestos.

And not only that, but we were taking it home to our wives and children without knowing it. We were carrying it home in our clothes and on our shoes and no one knew about it at the time. My first wife used to wash my clothes, and she died of lung problems. Makes you wonder how many other people have died like that over the years.

So now I've got it. I tire easy, I have high blood pressure, my intestines are all messed up, and lots of the time I can't sleep, I've got to sit up all night. Otherwise it's like you're drowning in the fluids in your lungs. You're slowly suffocating. There's no cure for it, no help, no medicine. You go to a doctor and he'll just say, "You've got to learn to live with it." I always think, "You mean I've got to learn to die with it."

*The major U.S. manufacturer of asbestos products.

Another problem the asbestos worker has had, which we're just now getting solved here in California, is that the insurance companies would stall you forever. They know you're going to die so they just try to wait you out. They deliberately stall and stall and stall. I filed my claim in 1970 and it took four years to settle it. They kept sending me to doctors, and then more doctors, and then even more doctors, and every time I'd call to find out what was happening, they'd always say the same thing: "It's being processed." Processed, processed, that's the magic word. The truth is, they're trying to starve you out, wait until you're so broke and desperate you'll settle for anything, whatever they feel like offering you, no matter how cheap it is.

Now in California we have this new law, AB946, that sets up a $2 million revolving fund, so that when it's medically proven that you have disabling asbestosis, from then on the fund will pay you so much to keep you going until they can collect from the insurance companies to pay back into the fund. So it doesn't cost the taxpayers a dime. Everything that gets paid out to disabled workers comes back in from the insurance companies. Because they know if you have asbestosis, they're going to have to pay your claim someday, but the way it used to work, they'd stall you for years, and at the same time they had that money out working for them, so by the time they finally got around to paying you off, nine times out of ten they'd already earned back from that money whatever they were going to pay you at the settlement.

These days asbestos is out of the new insulation they're putting in places. It's all fiberglass and rockwool now.* But people think that means the problems are over and it doesn't. Nobody's safe from it. We've got buildings and factories all over this country that are just loaded with asbestos in the walls, around the pipes, in the ceiling, on the floors, And when they get remodeled or demolished, look out, that's when you've got to rip all that old asbestos out, and unless you know how to handle it, it'll fly all over the place and contaminate the whole neighborhood. It's even more dangerous than when it was put in because it's drier and the air carries it easier. It'll travel further. People think they can stand across the street and watch it and they can't, unless it's being handled right. It's too dangerous otherwise.

*There is, of course, great concern that these substances may also be hazardous and they are being carefully watched. So far it appears that they only cause annoying skin irritation.

And I know for a fact that it's not being done right in a lot of cases. You look at some of these bids that subcontractors put in for removing asbestos on a renovation job and if you know anything about how it's supposed to be done safely, you'll know that there's no way they could even begin to bring in the right kind of equipment for the kind of bids they're submitting. It can't be done. There's a company down in southern California that is notorious for hiring wetbacks, paying them three dollars an hour to take out asbestos, and then firing them and sending them back across the border. But they don't tell them it's asbestos. They say it's something else, fiberglass or something. But it's asbestos. I've seen some of these jobsites and I can tell you for a fact that those poor guys are getting a hell of a dose of asbestos, so are all the other trades working on those jobs too, and so are all the people living in the neighborhood. Nobody's safe.

Then, on top of all that, you still find companies that have been taking the old asbestos that they rip out and dumping it in vacant lots. It's supposed to be put in specially tough plastic bags, marked with warning labels, and then hauled off to a site licensed to handle hazardous wastes, but some of these guys'll just pile it up anywhere they can find an empty field or a vacant lot. And you know what happens then. Kids find out about it and come along and play in it. So nobody's safe, as long as that kind of thing is going on. And there are other things, too, like brake shops, the way they blow that stuff all over the place when they work on your brakes, that's crazy.

You ask anybody that ever worked with asbestos if they had known then what they know about it now, would they still have worked with it, and they'll say, "No way." But the hell of it is, the companies knew all along, as far back as the nineteen twenties and thirties, that it could be harmful. But they lied to us. I for one would sure have picked another trade.

I love life and I'm going to hang on as long as I can, but it's chopping me down a little more each day. I'm slipping all the time. All my friends are dying around me. A good friend of mine, one of the nicest fellows you could ever work with, he died of asbestosis just recently. I went up to see him at the hospital the night before he passed away and he was just skull and bones, like a skeleton wrapped with skin. It's not right, just not right. The worst thing is that the companies knew all the time, but they didn't tell us. It's just not right.

Today we know that a high percentage, as many as 20 to 40 percent, of workers with a heavy exposure to asbestos will develop some degree of asbestosis. We also know that heavy exposure to asbestos alone increases the risk of contracting lung cancer, which is responsible for some 20 percent of the deaths in workers heavily exposed to asbestos. We know that if they smoke, asbestos workers are 80 to 95 times as likely to contract lung cancer as nonsmokers who don't work around asbestos. We know that workers exposed to asbestos have about twice the rate of gastrointestinal cancer as comparable populations of nonexposed individuals. We know that mesothelioma, which is invariably fatal within a year after its symptoms first appear, kills 10 percent of all heavily exposed asbestos workers, is on the rise near shipyards, and is a hazard even to those with very limited exposures, including in some cases those members

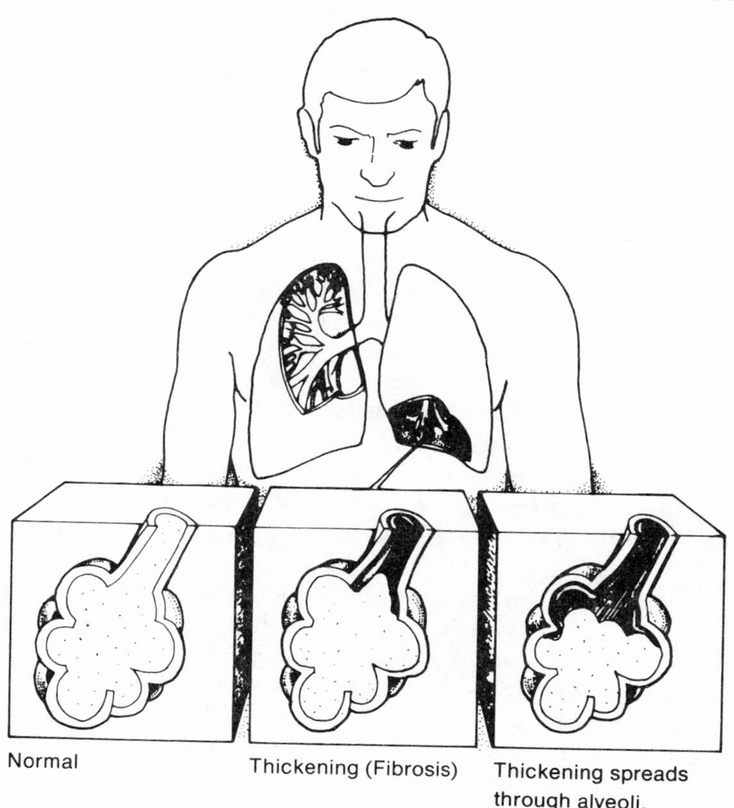

Figure 8

of a worker's family whose only exposure is to the dust carried home in an asbestos worker's shoes and clothes.

We know that the lungs can expel a lot of the asbestos that they inhale, but that once it has penetrated deeply into the alveoli, and the body has reacted to it in certain ways, it tends to stay, particularly the smaller, invisible particles that are the most dangerous. It does not, as workers were once commonly assured, "dissolve." We know that heavier, longer exposures, particularly for those who smoke, are the most dangerous, but we also know that even short exposures—as short as one week—can also be hazardous, even for nonsmokers. We know that the asbestosis associated with heavy exposure is a type of scarring of the lung alveolar air sac that blocks passage of air from the lungs into the bloodstream, and that this scarring causes an increased cancer risk.

We know that diseases caused by exposure to asbestos take a very long time, anywhere from ten to forty years, to appear in their victims, and that there is an enormous group of people with latent asbestos-related disease in this country. The leading national authority on asbestos disease, Dr. Irving J. Selikoff, has estimated that the final asbestos death toll among ex-World War II shipyard workers alone may reach 1 million by the end of the century.

Finally, we also know that living near a shipyard or other industrial operation that uses significant amounts of asbestos, or living in the same house with an asbestos worker who either doesn't know about or doesn't bother to follow safety precautions, such as leaving contaminated clothes and shoes at the work site, greatly increases the risk of contracting asbestos disease even for individuals who never set foot inside a contaminated workplace. One study of relatives of ex-asbestos workers found that more than one-third of them had X ray abnormalities indicative of asbestos exposure. The same study also disclosed the presence of asbestos fibers in the homes of some of these families *twenty years* after the plant they had come from had been shut down.

According to Ed Storey, Business Agent of Local 16, 57 percent of the deaths in George Fuqua's union from 1967 to 1975 were due to asbestosis and asbestos-related cancer. Storey is particularly worried about mesothelioma—of the last thirty-two cancer deaths that had occurred in Local 16 when we talked to him, 25 percent had been caused by mesothelioma, an extremely high percentage. And since mesothelioma can be contracted through very brief exposures to asbestos, he's especially concerned about the threat to members of asbestos workers' families, including his own.

I have a daughter who's twenty-two and a son who's seventeen. Both were exposed to asbestos when I worked in the field, until my daughter was six. That's the type of thing that you're really concerned about—your wife and your children and knowing that they were exposed. You know that you worked with this material, that when you brought your clothes home, the dust factor—that's the kind of thing that's constantly in the back of your mind, that something might happen to them. It's a concern you live with.

We had one family down in the San Jose area, the husband had asbestosis and a heart problem. He died two or three years back of a heart attack, but within one year his wife died of mesothelioma. And now of course both his daughters are very worried.

Neither Ed Storey nor George Fuqua believes industry will voluntarily control hazards like asbestos in the absence of effective monitoring and enforcement by watchdog agencies like OSHA (Occupational Safety and Health Administration). Storey puts it this way:

In a lot of areas industry has cleaned up its act, but only because of pressure and publicity. Without pressure being brought to bear, they wouldn't be responsive at all. They will, when they can, claim ignorance, especially on rip-out jobs. They'll claim they didn't know it was asbestos. In case after case that we see they'll send a worker in there that they pick up off the street, to rip out asbestos, and they'll tell him it's anything but asbestos. He doesn't know any different, but they do.

Industry is not going to safeguard worker health unless it's forced to. I don't know how to say it any plainer than that, because we'd all like to feel that industry is concerned, and that industry is going to do something, but it just doesn't seem to work out that way. Basically it boils down to money. They're aware of the hazards, and that it's expensive to do rip-out safely, but they'll turn right around and look for a cheaper way to do it anyway.

Fuqua echoes these views:

OSHA was having some trouble a little while ago because its inspectors were really getting pushed around and their morale was sinking. So they had a conference at the University of

California and they asked me to speak. I told them, "Sure, the last thing any employer or manufacturer wants to see is an OSHA inspector. You will never be welcomed in any of these companies," I said. "They don't want to see you. But OSHA is a lifesaver. If it wasn't for organizations like OSHA, the employers would just slaughter the workers."

Asbestos hazards may be at their worst in the workplace, but they are by no means confined to it. Asbestos is ubiquitous. It is in our food, our water, our beverages, in certain types of drugs, in a number of common household furnishings and small appliances, and above all, in our air.

Duluth, Minnesota, draws its drinking water from Lake Superior. Twenty-five years ago there was only a trace of asbestos in samples of that water. In 1974, however, levels averaging from 1 to 30 million fibers per liter, a very high count, were found.[3] The source turned out to be residues of an iron oxide ore, taconite, contaminated with asbestos that had for years been dumped into the lake by a nearby firm, the Reserve Mining Company.

No one yet knows whether this represents a hazard or not—or if it does, how much of a hazard it is. Studies of Duluth cancer death rates for the 1950s and 1960s found some increase in mortalities from rectal cancer, pancreatic cancer, and GI tract cancer in both sexes, as well as increased stomach, small intestine, and peritoneal cancer in men, but investigators concluded that not enough time had passed to assess accurately the true impact of asbestos in Duluth's water supply.[4]

Other areas besides Duluth are also affected. A 1980 study[5] of seventy-four drinking water sources in the San Francisco Bay Area, many of which contain asbestos derived from the serpentine rock common to aquifers and reservoirs in the area, revealed a significant relationship between asbestos levels in these sources and the incidence of lung cancer in white males, gall bladder and pancreatic cancer in white females, and peritoneal and stomach cancer in both groups.

While findings like these are suggestive, it is premature to conclude that asbestos in drinking water causes cancer. Additional studies in which other factors that are known to contribute to elevated cancer rates, such as smoking, diet, and occupation, are considered and carefully controlled for, will need to be conducted before we can say with certainty whether asbestos in our water supplies is carcinogenic, and at what concentrations it poses this hazard.

Other beverages besides water are also suspect. In 1971 a survey of two dozen samples of Canadian and American tap water and a variety of beers, wines, sherries, vermouths, and ports found asbestos fibers in all of them, ranging from 1 to 4 million fibers per liter for some of the beers to as high as 12 million fibers per liter for some soft drinks. Not surprisingly, the highest count, 172 million fibers per liter, was found in a sample of tap water taken from a lake in the heart of Quebec's asbestos mining district.

We are also exposed to asbestos in consumer products and appliances. In 1980, alarmed by the amount of asbestos being used in washers, dryers, electric mixers, refrigerators, and a host of other consumer products and household appliances, the Consumer Products Safety Commission moved to gather information from firms selling or importing products containing asbestos on details of its use in these products. This followed an earlier order in 1979 in which 22 million hair dryers manufactured by some forty different companies were recalled because of concern on the commission's part that they were emitting respirable particles of asbestos.

Even the typical homeowner who does some of his or her own remodeling may be exposed to elevated levels of asbestos released from construction materials. For example, one of the more worrisome uses of asbestos uncovered by the Consumer Products Safety Commission in its 1980 survey was in an asbestos shelf-lining paper, sold casually in hardware stores and recommended for use near stoves, heaters, and other heat sources, that emitted clouds of respirable particles when cut or bent. Another example: Asphalt-vinyl tiles, installed on countless floors over the last few decades, are still widely used and contain anywhere from 15 to 25 percent asbestos. In some cases, preparing to resurface one of these floors when it gets worn involves sanding the old tiles so that the adhesive cement under the new floor will have a rough surface to adhere to. Sanding can release hazardous amounts of respirable fibers into the air at levels comparable to those released in shipyard work, and is known to have caused asbestosis and mesothelioma in men who regularly worked as floor tile installers.[6]

But by far the most hazardous source of asbestos exposure is the air we breathe. Asbestos fibers are found in air samples from all around the earth. All of us, especially urban dwellers, have some degree of asbestos contamination in our lungs, though normally at much lower levels than in those of individuals exposed to asbestos in the workplace.

Nonetheless, episodes of heavy contamination of the general environment are not unknown. Until 1973, when the practice was first banned by the EPA, it was common in urban areas for construction crews working on high-rise structures to spray the steel columns and beams with a fire retardant material containing asbestos. On windy days a large burden of asbestos was dispersed for miles around in this way. This practice was first halted in Manhattan in 1972 when Dr. Selikoff himself measured air concentrations at City Hall and found asbestos at dangerously high levels. In many cases the inside of ductwork through which air is circulated in high-rise buildings was also sprayed with a fire retardant material containing asbestos, which means that as these buildings age their interior atmosphere may become increasingly polluted with respirable asbestos fibers.

In June 1977 Dr. Selikoff and two colleagues reported[7] that ore containing asbestos was being mined in a quarry in Rockville, Maryland, crushed, and then used throughout the locality as surfacing on roads, parking lots, driveways, paths, and playgrounds. Measurements of the dust raised by traffic on roads surfaced with this material revealed an asbestos content one thousand times the airborne levels found in the average U.S. city, and well above levels known to be capable of causing disease. The authors urged the EPA to declare the quarry, which had been in operation for twenty years, a public health hazard and to ban further use of its materials. Within two months it was shut down.

As more and more understanding of the hazards of asbestos inhalation has developed, federal exposure standards have steadily dropped. As we saw above, in the late 1930s the U.S. Public Health Service advised the equivalent of about 170 fibers per cubic centimeter (cc). In 1972 OSHA set an emergency standard of 5 fibers per cc for fibers less than 5 microns in length, and in 1976, reduced it to 2 fibers per cc, for fibers under 5 microns long. (Generally, fibers below 5 microns in length are considered to be the most highly respirable, and thus the most dangerous. But, depending on their shape, fibers up to 15 microns in length may also be highly respirable.)

Many experts now feel that the current standard is still too high, and should be lowered to ½ fiber or even 0 fibers per cc. They point out that the current standard of 2 fibers per cc still allows someone breathing a normal 2.4 cubic meters of air in an eight-hour day to inhale 4.8 million fibers, at rest. If he or she is at work, that figure rises by a factor of 10, to about 50 million fibers a day. As Dr. Selikoff has pointed out, "There's a lot of published evidence that in two fibers per cc mesothelioma occurs, lung cancer occurs, asbestosis occurs."[8]

But in dealing with a known human carcinogen, like asbestos, there is often no such thing as a "safe" exposure level. The real issue is money. In many industries the asbestos exposure level could be lowered to ½ fiber per cc of air with relatively small expense, but as OSHA itself acknowledges, in many others the cost would be so high that these firms would in effect be driven out of business, and that would of course mean the loss of irreplaceable jobs. So, as in any other aspect of environmental health, interdependence and its corollary, trade-offs, emerge once more as fundamental principles: setting safety standards, we see, is inevitably as much a function of economic realities as it is of medical concerns.

As far as the exposure hazards of the general public go, Dr. Selikoff believes that "the possible prevalence of effects has not been adequately studied; too often environmental scientists have been confronted with the statement that 'there is no evidence of any harmful effect' when, in fact, no one had really looked. It may be true that there is no threat, but until it has been demonstrated to be true one cannot close one's eyes to the possibility."[9]

DBCP

Since World War II there has been a massive proliferation in the use of synthetic pesticides in this country, one that we discuss further in Chapter 6. Today, some one thousand chemical compounds are registered as pesticides in the United States, and some thirty-five thousand different products in which they are used in different concentrations and formulations are on the market for everyone from the conglomerate that farms thousands of acres of Midwestern wheatland to the homeowner who buys Triox to kill his weeds and pentachlorophenol to preserve his fence posts. Until the late 1970s a pesticide* that was very popular with farmers who had to worry about nematodes was DBCP (1,2-dibromo-3-chloropropane).

Nematodes are a species of parasitic roundworm that lives in the soil and attacks a wide variety of crops, including citrus, grapes,

*The term "pesticide" includes insecticides, fungicides, herbicides, fumigants, and rodenticides, among others.

peaches, pineapples, soybeans, tomatoes, nuts, and bananas. DBCP is a very effective nematocide with some unique properties: Once injected into the ground, it remains active for several years without reapplication, yet does not contaminate or harm plants grown in the treated soil, especially perennials like citrus and grapes. Other nematocides not only kill the worms but also everything else growing in the soil.

Because pesticides are deliberately designed to kill living organisms, government agencies responsible for safeguarding public health have traditionally tended to pay special attention to their potential to cause disease. However, until very recently this concern focused mainly on acute effects. Long-term toxic effects from relatively low, continuous doses were not considered to be as significant a problem. However, events of the last decade have changed this attitude. We now realize, for example, that many cancers are caused by chronic exposure to low concentrations of substances, concentrations well below the threshold at which acute effects appear. We have also, due in part to the pioneering work of Rachel Carson and others, become slowly but steadily more concerned about the potential of many pesticides to interfere with reproduction. *Silent Spring,* published in 1962, pointed out that DDT was having adverse effects on the reproductive cycles of a number of different bird species, and even presciently hinted that this effect might appear in humans, too: "Some indication of the possible effect on human beings is seen in medical reports of oligospermia, or reduced production of spermatozoa, among aviation crop dusters applying DDT."[10]

Four years before Miss Carson's prophetic comment was published, a University of California researcher under contract to the Shell Oil Company had tested the effects of DBCP on lab animals and identified an exposure level that appeared to be safe. He had noticed that the compound caused atrophied testicles in male animals, but attached little significance to this finding at the time. The most important effect, it seemed, was that DBCP could cause liver and kidney damage at concentrations higher than those capable of causing testicular atrophy.

In 1961, in conjunction with the Dow Chemical Company, which had independently conducted its own tests on DBCP and arrived at the same conclusions, this researcher's findings were published in the journal *Toxicology and Applied Pharmacology.* The U.S. Department of Agriculture, which was at that time in charge of monitoring and enforcing pesticide safety standards, reviewed the test results and concurred that despite the evidence of testicular atrophy in lab

Figure 9

animals, a safe level of exposure for humans making the compound and applying it could be established. DBCP was assigned a class III designation, the least hazardous category, by the USDA.

For the next two decades DBCP was widely formulated, sold, and applied by farmers all over the country. In 1973 the National Cancer Institute published data indicating that DBCP was carcinogenic in lab animals, and the EPA, which had taken over the job of regulating pesticides from the USDA in 1970, slowly began to review the DBCP data to see if regulations governing its use should be tightened up. In 1975 new information showing that DBCP was also mutagenic came to light. But despite the earlier reports on DBCP's ability to cause atrophied testicles in test animals, until well into the 1970s it was widely believed that the only pesticide capable of causing sterility in human males was kepone, a substance used in ant control.

One place where production and sale of DBCP had been taking place since 1962 was a chemical plant in Lathrop, California, owned by the Occidental Chemical Company, a subsidiary of the Occidental Petroleum Company. By the early 1970s the plant was producing one thousand to twenty-five hundred tons of DBCP a year. None of the workers who came into contact with it every day were ever warned that it might be hazardous, and consequently they took virtually no safety precautions whatsoever to protect themselves from it. Today we know that the judgments made by industry and federal officials in the 1950s to 1970s were wrong: a number of the Lathrop workers were sterilized by contact with DBCP, some permanently. One of them was Teddy Bricker:

> I started out with Oxychem [Occidental Chemical Company] fourteen years ago, in 1967, in the lawn and garden section, which is a small department in the warehouse. I worked there for about a year and a half, then I went to Agchem, the Agricultural Chemicals Department, which was where we formulated and blended DBCP and other insecticides, fungicides, herbicides. Our job was to take the concentrated raw materials and blend them into the right formulations for farm use. One reason the farmers today are still saying, "Hey let us keep using those chemicals," and the chemical worker is saying, "Wait a minute, we need more protection," is that they don't understand that the concentrations we have to work with are much higher than what they're using out in the fields.
>
> I worked with DBCP from 1971 to 1977. At first we treated

it just like it was water or any other nontoxic material. There weren't any precautions like wearing a respirator—as far as we knew it was harmless.

We would mix DBCP by putting allyl chloride in the reactor, then we shot it with bromine to brominate it. We'd watch the temperature and the specific gravity to make sure we didn't overbrominate it, then when it got to a certain point we'd put it in a drying tank. There it recirculated through a filter system, through some nylon screens, and then it passed through some catalyst balls to help take the moisture out of it. Finally we'd put calcium chlorate on top of it to take up any leftover moisture, then we'd scrape that off, dump it on the ground, and wash it down the drain.

They figure now that it was in the filtering stage and later in the process of putting the finished DBCP in cans to ship it out that we got our worst exposures. When those filters would clog up we'd go and spray them out with water, to open them up. Sometimes we'd use gloves, but the majority of the guys didn't, or if they did, it was mostly to keep the smell of the bromine off their hands.

There's a chemical reaction in the bromine process that makes it heat up. It creates its own heat. So you'd also get exposed to the vapors coming off there, too.

At the end of the formulation process, we'd add EPC [epichlorohydrin] to take the acid out. And that would be the end of the process. The finished product was DBCP, which is a liquid nematocide.

We wore no protective clothing at the time, except for maybe glasses, a hard hat, and rubber boots. Coveralls, a hard hat, and glasses was our basic work uniform at the time. And the only reason we wore rubber boots was EPC could burn a blister on you. So we'd wear boots if we had them. If we didn't, or thought we could handle the EPC without getting in it, well then we didn't wear them. Some days were worse than others. Sometimes the process would go along real smoothly, other times we'd have our problems.

When we canned the DBCP, we took it from the storage tanks, put it in a mix tank, and added solvents and emulsifiers to make either DBCP 12 or DBCP 12E. Then we would can it out, seal up the cans, put them on pallets, and ship them out. There was no ventilation system to speak of, so we had toxic exposures there, too. If we had a spill when we filled up the cans, we'd wipe it up with a rag, then throw the rag away. We were doing that whole process, from raw ingredients to the final formulation that we sold to farmers, with virtually no protection.

When I started out with Agchem in 1969, it had about ten people. In the busy season we'd pick up more people because Occidental would have a backlog of contracts and we'd have deadlines to get the stuff shipped out. We were often working on a rotating shift, so that work was going on around the clock and we'd have to work at all hours of the day or night. During those busy seasons we would have as many as twenty to twenty-five people working in the Agchem department. Then during the regular season we'd drop down to ten or maybe twelve at the most.

We ran a lot of other things, too, that at the time they said were safe to work with, things like heptachlor, chlordane, aldrin, dieldrin, TEPP, DDT, things that have since then been taken off the market because they're all considered carcinogenic. This was from 1968 to, oh, maybe 1975, or even later for some of them.

People tell me now, "You ought to have known better—you're working in a chemical plant; of course that stuff's dangerous." But they forget that those things are only found out bit by bit as time progresses. Who knows, something that we're working with today, next year they might find that it's carcinogenic and take it off the market, too. EDB [ethylene dibromide] or something like that, something else that we're still making here. People don't realize that.

Back then, we were told that if we got poisoned they would give us atropine to counteract it, and we could go on about our business. And that's the way they treated poisonings. Unless you were deathly ill, or thought you were deathly ill, you'd very seldom ever go to the hospital. You might go to the doctor's office, he'd give you some atropine, you'd either go home from work that day or you'd go back to work, finish your shift, go home and take it easy for the rest of the day, and then come right back to work the next day. It was treated just like the common cold. If some guy got sick, they'd figure, "Well he's just coming down with the flu, or he's just having a bad day, or maybe he ate something bad," and they'd send him home. But as soon as he came back to work the next day they'd put him right back in the same place with the same exposure to the same chemicals.

About three or four years after I started working for Oxychem, guys who worked in the Agchem department were beginning to wonder whether something was wrong with their ability to have children. But they weren't really taking it too seriously. They used to joke with new guys in the department, saying, "Hey if you want to have children you better not work here." But they had no proof and everyone just laughed and didn't give it a second thought. They couldn't go to the company with a serious complaint and say, "We've got a serious problem here," because they had no proof.

But then one of the guys in the department, he and his wife really wanted to have children but they couldn't. She went to the doctor, they ran some tests, and said everything was normal with her, she should have no problems conceiving and should have a normal pregnancy. Then he went and got tested, and came back with a long face and told Jack Hodges, our chief union steward at the time, that the doctor had found out he was sterile.

I got involved with the whole thing at that point because I was on the Health and Safety Committee, so Jack came to me and we talked about it. Then both of us met with Rex Cook, who was the secretary-treasurer of the union at the time [Oil, Chemical, and Atomic Workers Local 1–5] and we talked about it some more. Then I brought it up at several different meetings of the Health and Safety Committee, but the company didn't do anything about it.

Of course when the other guys working in the Agchem department heard that one guy actually was sterile, they all assumed they might be, too, even though his doctor hadn't specifically said anything to him about the possibility that it might be the chemicals he was working with. As we all talked it over, it began to dawn on us that a lot of us hadn't been able to have children for quite a while, several of the ten or so regulars in the Agchem department. They'd been trying, but no children.

I myself had one son in 1969 and another in 1973. After 1973 I worked with DBCP a good part of the time, and that was probably when my own sperm count started to drop.

So Jack and I got together and decided we had to come up with some proof or the company wouldn't act. We needed evidence. At about the same time, a couple of filmmakers, Josh Hanig and David Davis, were making a film about the dangers of the American workplace called *Song of the Canary,* and they'd asked Rex if he could help them.* So he had the idea of telling them they could film some of us if they'd pay for our sterility tests. We knew we had to do something—it was just a matter of figuring out how to get the guys to cooperate and give samples—but in the process Josh and Dave and another fellow named Steve Moser just happened to come along. It seemed like a prime opportunity to get some of the testing work we needed to have done paid for, so that's what we did. We asked them and they agreed.

The next step was to get the guys in the Agchem department to go along with being tested. At first, when they heard what the process was, they said, "You're crazy, who do you think you're going to get to—" Well, it was just the fact that the testing process is pretty embarrassing for some people. It can be pretty hard to ask a grown man to do stuff like that.†

So, at first the guys said, "You're nuts." And we said, "We have to have some proof—the testing will be done in a medical manner, in privacy, with doctor-patient confidentiality, and all that." Finally we convinced them, seven of them, I think it was. We had about ten at first, but only about seven actually went through with it. Then we also had a couple of guys who had had vasectomies offer to help. They said, "If it'll help, I'll volunteer, I'll go ahead and get tested, too." So we finally wound up with seven samples. And out of the seven, four were completely zero, no sperm count at all, and the other three were low. So we had the proof we were looking for. We took the evidence to Rex and that's when he called Dr. Whorton. At that point we had something to go on. That's when the public health department was first notified that we had a problem out here.

After these initial test results were confirmed by a second medical evaluation, conducted by Dr. Whorton, the workers' union and the management of the Oxychem plant agreed to allow him to examine all thirty-six male and three female workers in the Agchem division at that time. Of these thirty-six males, eleven were found to have virtually nonexistent sperm counts (i.e., below 1 million per milliliter), three had low counts (10 to 30 million/ml), and eleven had normal counts (above 40 million/ml). (The remaining eleven male

* New Day Films, 1454 Sixth Street, Berkeley, California 94710.

† The test requires either coitus interruptus or, preferably, masturbation into a semen sample collection bottle.

Agchem workers were excluded from the evaluation because they had all had vasectomies.)

Even more alarming than the low sperm counts was the clear connection found between the length of time a man had been working with agricultural chemicals in the plant and the extent of damage to his sperm production system. Every worker who had been employed in the Agchem department for at least three years had a sperm count below 1 million/ml. No worker with a normal count had been exposed to DBCP for more than three months. Those in the intermediate range, with counts from 10 to 30 million/ml, had been exposed to DBCP for about a year.

Analyses of samples taken surgically from the testicles of ten volunteers confirmed the nearly complete absence of sperm in men with long, heavy DBCP exposure, and revealed signs of degeneration in sperm-producing cells and conducting passages identical to those observed in the earlier Shell-Dow studies.

At this point there was no question that something was seriously wrong at the Lathrop plant. Although there was very little doubt in anyone's mind that the sterility was caused by exposure to chemicals at the plant, no one knew precisely which ones were to blame. Toxicology data on the hundred or so chemicals used in the plant revealed that not just one but *four* were known to have the potential to cause damage to male reproductive systems in laboratory animals. But further research led to the conclusion that of the four, only one, DBCP, was present in quantities large enough to have caused the extensive effects seen in the men tested.

Concerned about the scope of the sterility problem revealed in the first two screenings, the supervising physicians recommended that all men working at the Lathrop plant be checked for possible reproductive effects. Both the union and the plant management readily concurred. A total of 196 men were examined and asked to fill out questionnaires. Evaluation of the results showed that of all the men exposed in any way to DBCP, 13 percent had virtually no sperm count at all, 17 percent had a severely lowered count, and 16 percent had a count in the low normal range (20 to 39 million/ml). Almost half of the men in the plant with any exposure to DBCP at all, not just those who worked in the Agchem department, had suffered some degree of testicular dysfunction.

Publicity over the Lathrop findings prompted Dow and Shell to survey workers at two other plants, in Arkansas and Alabama, where DBCP was also manufactured. In the Dow plant in Arkansas, of the first fourteen workers examined, twelve had no sperm count at all. In all a total of some one hundred men damaged by DBCP exposure were identified by these and other studies in 1977–78. The only chemical that workers were exposed to in all three plants—in California, Arkansas, and Alabama—was DBCP. In August 1977 both Dow and Shell, seeing the handwriting on the wall, stopped producing DBCP and recalled all their outstanding stock. There is no doubt that they were motivated in large part by genuine concern for workers in their plants, but there was another significant factor: money. The patent on DBCP had expired in 1977, and since it is not a difficult or expensive compound to formulate, loss of patent protection meant that these companies were now vulnerable to competition. As well as the moral and legal issues raised by the revelation of DBCP-related injuries among their employees, manufacturers of DBCP also faced a tough economic question: Would they be able to continue to make DBCP profitably? Evidently the answer was no, since all three—Dow, Shell, and Occidental—decided to halt production permanently. However, a smaller Los Angeles company soon stepped into the gap they left and began making DBCP, primarily for export.

In September 1977 Cal-OSHA, the state counterpart to the federal agency, recommended temporary closure of the Lathrop plant, which voluntarily complied. The California State Department of Food and Agriculture banned DBCP use throughout the state. Federal OSHA issued a temporary emergency standard for DBCP, limiting its allowable concentration in the workplace to 10 ppb, thought to be the lowest level that could be measured at the time. (OSHA is not permitted by law to ban hazardous substances, but it can achieve essentially the same result in many cases by setting allowable exposure limits so low that employers find it economically or technologically impossible to comply. In 1977 it was OSHA policy to attempt to do just this for any workplace hazard found to be carcinogenic.)

By December 1977 OSHA was holding hearings to determine a permanent exposure level for DBCP. Three months later, having learned that it was possible to measure concentrations as low as 1 ppb, it established this level as the permanent standard. If it had been possible to measure concentrations as low as 1 part per *trillion,* that's what the OSHA standard would have been. In that same year, 1978, six more studies of human exposure to DBCP confirmed the Lathrop findings.

Slightly earlier, at about the same time the OSHA hearings were being held, the EPA had issued a number of restrictions sharply limiting the use of DBCP on most fruits and vegetables, but permitting the continued use of some 4.5 million kilograms a year on certain perennial crops, namely peaches, citrus, grapes, nuts, and pineapples. Not until nearly two years later, in August 1979, did the EPA convene hearings on the matter of permanently canceling DBCP's registration, in effect banning its use completely. After hearing the testimony offered at these hearings, the presiding judge recommended that DBCP be, in essence, banned.

In response to this recommendation, two of the largest users of DBCP, Del Monte and Dole, threatened to withdraw from the pineapple business in Hawaii and invest more heavily in their pineapple plantations in Taiwan and the Philippines. The loss of jobs that this move would have caused so terrified the Hawaiian pineapple workers union that it endorsed the request by Del Monte and Dole for an exemption from the proposed ban. On October 29, 1979, EPA Administrator Douglas Costle bowed to this pressure and issued a suspension order that prohibited all uses of DBCP except on pineapples in Hawaii.

Since DBCP was widely used in other countries, the findings at Lathrop and shortly after in Arkansas and Alabama set off international alarm bells. Soon studies of field workers in countries like Costa Rica and formulators in Mexico and Israel had turned up the same pattern of increasing testicular damage with larger, heavier exposures.

In 1980 the first evidence of possible DBCP effects in women was reported in an Israeli medical journal.[11] The sixty-two women studied were married to men working on banana plantations in the Jordan valley. The use of DBCP had been banned in Israel in 1977, almost as soon as the news of the Lathrop findings had been reported. But an exception had been made for those particular banana plantations because nematode infestation was particularly bad in the Jordan valley. Stringent safety precautions had been followed, including the use of protective clothing and respirators, to protect DBCP applicators on these plantations, but there were indications that their reproductive systems were being damaged anyway.

The 1980 study involved interviews with the workers and their wives, questionnaires, and careful reviews of hospital pregnancy records. It showed that before their husbands had come into contact with DBCP, the rate of spontaneous abortion in this group of women was only 6.6 percent. After contact, however, it rose to 19.8 percent. There seemed to be little doubt that DBCP exposure in these men was related to increased rates of spontaneous abortion in their wives.

Ironically, the earlier Shell-Dow research on DBCP, whose hints of testicular damage had been overlooked for two decades, now suggested a hopeful possibility: It had been observed in the original studies that the testicular damage observed in lab animals exposed to DBCP was occasionally reversible. In certain cases damaged sperm cells and other organs, including kidneys and livers, had regenerated. Tests made on the Lathrop workers one year after their exposure to DBCP ended were inconclusive. Of the twelve who had suffered total sterilization, none had recovered. However, of the nine men with shorter exposures and abnormally low sperm cell counts, six had fully recovered. The results suggested that if DBCP exposure is stopped soon enough, before it has caused complete degeneration of sperm production cells, damaged cells may recoup within twelve to sixteen months. The same finding has been shown by other investigators in the United States and Israel.

The question of who is responsible for the Lathrop incident is now being settled in court. Lawsuits claiming millions of dollars in damages have been brought by Workers' Compensation attorneys against Oxychem and by numerous third parties against Dow, Shell, the University of California, and the U.C. researcher who conducted the original toxicology testing for Shell.

One suit brought by the EPA and California against Occidental was settled in early 1981. Under its terms Occidental has agreed to pay $2.7 million to support environmental health research and to cover some of the costs of evaluating and correcting the environmental damage caused at the plant by its indiscriminate dumping of DBCP in the area. It has also agreed to implement an extensive, twenty-year cleanup program at the plant site that will cost another $15 million.

According to Teddy Bricker, the larger questions of responsibility will never be settled:

> Since the story has been publicized, a lot of people have asked me, "Well if you thought you had a problem, why didn't you do something about it?" We did. But they forget that it takes time to organize something like this. And until the media got ahold of the story, we also had to be careful to keep everything confidential as possible. We had to be careful about who had access to the information, since it could have been embarrassing to some of the workers. We still try to be careful about that.
>
> Furthermore, at no time did we get any kind of help from the company. At no time did the company, which knew all about the situation from the beginning, say to us, "Well, if you guys

think you really have a serious problem there we'll give you the benefit of the doubt and help you, we'll help you run the tests if you will participate in them." They never offered. When we brought the problem up the first few times in health and safety meetings or in contract negotiations, they just ignored the whole thing. That's really where the company let us down. We didn't know it was specifically the DBCP that was causing this problem at the time, but we had a pretty good idea it was one of the chemicals in our work site, what with four out of seven of the sperm samples coming up zero. We knew we had a problem but even then, even after the proof, they never offered to help us set up some kind of a program to help us out. We had to work it out ourselves. And when you're talking about working and trying to get a group of other working people together to coordinate something like this, and about getting funds to pay for the tests, and about getting testing schedules set up, you're talking about something that takes time. It didn't just suddenly show up overnight.

As of today [December 1980] the guys that had a zero sperm count back then haven't recovered. It hasn't reversed for them. Those whose sperm count was low were usually the ones who did just some of the canning but usually didn't have too much direct involvement with the actual formulation process. Some of them have sperm counts now that are normal or close to it. So it does seem to be reversible in some cases. In the last two years three of them I know of have fathered children.

My personal feelings today are that we were cheated all the way down the line, in a couple of different ways. First, we weren't ever told that the way we were making a living involved the possibility that we might not be able to have the families we wanted. I was lucky enough to have had two sons before I was sterilized. But some of the other guys weren't so lucky. And that was a decision that we didn't make ourselves. Now, if they had given us the alternatives to start with, spelled them out, and said, "Look you are working in conditions that might make you sterile, that might make it impossible for you to have children, ever, so you can either work here or not," I think most of the guys would have said no.

Second, we're finding out today that some of the original research that was done on DBCP, going back as far as 1958, shows effects like shrinking of the testicles in lab animals. Now, that's something that I think there should have been a lot more concern about before allowing humans to be exposed to the stuff.

Another thing is that we also know that DBCP and a lot of the other chemicals that we've been working with are carcinogenic. What's going to happen to us a year from now, or five years, or ten? And there's also the possibility of genetic mutations in some of our kids. What about them? Nobody really knows what they might be. That hurts, too.

I think everyone wants to know what their alternatives are before they make choices like these. Smoking is hazardous to your health, but at least the choice is ours. Whether it's a good choice or a bad choice, the right to make it is ours. We didn't have that right in the case of these chemicals. And it's not only DBCP—there's other chemicals, too, all over.

Now a lot of places where chemicals are used have monitoring systems and controlled atmospheres, but somebody had to pay the price, had to play guinea pig, and in our case we didn't have a choice about it. They've used us as guinea pigs, but didn't give us any alternative. We didn't make the decision, they made it for us.

There's another thing. There are still chemicals in the plant I work in that there is no medical research on. Some of the raw chemicals we receive are from foreign countries, mostly because they're cheaper. Some aren't even labeled in English, so you don't know what they are. I've seen burlap bags full of ziram from Germany, or Korea, and you can't read the label. Guys who work with it come up coughing mucus and blood and bleeding from the nose, but no one seems to have any idea what its long-term health effects are. This I'm against. I don't think anyone should have the right to jeopardize someone else's health or life if they don't have complete medical research on the effects of the substances they're using.

Also, the labels aren't always clear, even if they are in English. They don't always give antidotes, like "Go to a doctor immediately," or "Wash your hands for at least fifteen minutes." Then, too, here in California the majority of the farmworkers are Spanish-speaking and they can't always read a sign in English that says something like, COULD BE HAZARDOUS TO YOUR HEALTH. WASH YOUR HANDS. WEAR RUBBER GLOVES OR PROTECTIVE EQUIPMENT. KEEP OUT OF THE REACH OF CHILDREN.

Then there's just a lot of general irresponsibility around these days, too. You see a lot of that on farms out here. I've been out there and you'll see an applicator truck pull up, drop its hoses down into the nearest irrigation ditch to draw up the water they need, and the next you know the water in the ditch is turning green, yellow, blue, and every which color because of the chemical residues in that hose. Then that water floats on down to the next farmer's land where the cows are drinking from it or the farmer is spraying it on his crops. Most of these applicators couldn't care less. All they know is they're getting so many dollars a day to put that stuff on.

It's true that conditions are a lot better at Oxychem now than they were a few years ago. Since 1977 it's improved a lot. At times we still have a few messes in there, but not toxic exposure, or at least not so intensely, just general messes. I mean it *is* a fertilizer plant, so you're going to have dust, oversprays, slippery conditions, and messes from time to time, but overall we've improved it by 100 to 200 percent over what it used to be like. We have masks now, better ventilation, a decontamination area where we go in before we start the shift and change into our work clothes, and come back to on the contamination side and change out of them before we go to lunch or go home. And there's someone in there who washes the coveralls we wear, so we're putting on clean coveralls twice a day.

We have better waste handling procedures now, too. Empty chemical bags are compressed and hauled off to special dumps if they are from a toxic substance. Even if it's only suspected that the material may be toxic, we're supposed to treat it like a poison and wear a fresh-air mask when we handle it.

But then you still get supervisors who place the production schedule over the workers' safety. They'll rush you and say things like, "Don't worry about it, we'll take care of it, you're not going to be running all that much, let's just get it done."

And it's true that the equipment you have to wear to be safe isn't all that easy to use because it's uncomfortable. Some guys'll put it on until the supervisor leaves, then they'll take it off. And there are some supervisors who know the guys are doing that, who even watch them doing it, but who don't say anything about it because they just want to get their production run finished.

We also have better medical monitoring now, too. None of us knew anything about cholinesterase levels before, but we all do now.* Base-line levels are taken when a guy first comes to

*Cholinesterase (CHE) is an enzyme that helps regulate nerve transmission. A number of pesticides, especially the organophosphates, remove CHE from the body. This

work here, and they're monitored all the time. If we get sick we're pulled out and kept out until the readings are okay again. We know where to go if we get sick after work, and we know what the symptoms of toxic poisoning by these materials are like. But still, very seldom does a month go by without someone getting poisoned out here.

I've never had any problems of harassment or reprisals from the company. I don't know if I'd describe them as sympathetic or helpful—concerned about their liability is more like it. But I've spoken at many hearings and meetings on behalf of better safety conditions for chemical workers, and if they were going to badger me about it they would have done so long ago. But no one's ever threatened me, and things are better here than they used to be.

The lesson of all this is, if you don't fight for it, you won't get it. The company's not going to take money out of its profits to pay for safety, if they don't have to. They won't go for that. We didn't get any of these improvements without fighting for them. If I had the opportunity to address every union in the country, the only thing I would tell them is to fight for health and safety clauses in their binding agreements. If they don't have those they don't have anything. Otherwise, what the company gives you today it can take away tomorrow. But with health and safety articles in a contract you can build on it and have a better and safer place to work.

There are a number of other lessons that the DBCP story can teach us. First, we don't pay enough attention to the early warning signs of impending environmental health hazards, even when they are clearly spelled out in publicly available, scientific documents. Academic researchers, industrial interests, and federal agencies all had ample opportunity to prevent the damage that DBCP did at Lathrop and other places, and all failed. As happens far too often in industrial cultures, it was only after the damage had been done and people had been hurt that we did something about a problem that never should have been allowed to happen in the first place. Had there been in the late 1950s as much concern about the long-term, chronic effects of exposure to substances like DBCP as has now been aroused by the plight of men like Teddy Bricker, DBCP

---

allows nerves that regulate muscles to be overstimulated, leading in mild cases to flulike symptoms and in severe cases to death.

would never have been allowed into production without more stringent safeguards.

Now that our concern has been aroused by the DBCP episode, we are being more careful about substances that appear to have the potential to cause sterility and other reproductive disorders in humans. Workers exposed to these substances are being much more carefully monitored and trained in appropriate safety precautions. The possible reproductive hazards of tobacco smoke, marijuana, and heavy alcohol use are being explored. Some scientists, pointing to the sensitivity of the male testes to toxic chemicals, and to the tendency of toxic substances to accumulate in the male reproductive tract, are investigating the controversial theory that industrial contamination of the environment has caused a widespread drop in sperm count in the general male population over the last few decades.

However, despite our stepped-up activity in this area, it's a virtual certainty that in many cases our concern will be too late. Without a doubt the same mistake of waiting until the damage has already been done is going on right now for other substances, substances for which the harbingers of future disease outbreaks already lie present but unheeded in such places as scientific journals and in scattered clinical reports.

The final lesson that the DBCP story is teaching us is one that we know already but seem to need to keep relearning: the interdependence of the biosphere and of all its inhabitants means that environmental health hazards are never limited to the workplace. DBCP is unfortunately one of the few pesticides that is persistent enough to be absorbed into groundwater aquifers and to remain there for years at a time. One-third to one-half of the groundwater wells tested in California's major agricultural regions, the Central and San Joaquin valleys, now contain measurable levels of DBCP. Those with concentrations higher than 1 ppb have been closed to human consumption. In some wells in this area, DBCP levels as high as 5 ppb have persisted since 1977. Wells in other parts of the country where DBCP was heavily used, such as parts of Arizona, have been found to contain even higher levels of DBCP, ranging from 4.6 ppb to as much as 18.6 ppb. Traces of DBCP have also been found in drinking water sources in Los Angeles and in some wells in Hawaii.

Unfortunately, the popularity of DBCP among grape farmers in California has led to the birth of a black market in DBCP there. It remains to be seen how much of a hazard to groundwater supplies this lingering use will pose.

PCBs

The harbinger process can also work in reverse: a poisoning incident in the general public may reveal the toxic potential of a substance used in high concentrations in industrial settings, causing those who work with it to become justifiably concerned about the risks of their exposure to it. This is what has happened with PCBs.

PCBs (polychlorinated biphenyls) are a group of some 209 compounds—both solids and oils—that were first synthesized in the late nineteenth century and which entered commercial distribution in 1929. They are chemically inert and highly stable, which means they are very resistant to reaction with other chemicals or to natural processes of degradation. PCBs became popular because they help prevent electrical fires. Cheap, highly nonflammable, and very poor conductors of electricity—all these properties make PCBs invaluable cooling and insulating fluids in electrical equipment like transformers, capacitors, and electromagnets. (On the scale of flammability used by Underwriter's Laboratory, where water has a value of 1 and ether has a value of 100, PCBs are down in the 2–3 range. Few substitutes for them are quite so nonflammable and as inexpensive.)

Far and away the greatest use of PCBs since 1929 has been in electrical equipment, but other applications were quickly found as well, in hydraulic systems, compressors, heat exchangers, pesticides, paints, caulking, adhesives, carbonless copy paper, color television sets, microwave ovens, air conditioners, fluorescent light fixtures, ham radio sets, and even, it is rumored, in one brand of chewing gum, where they were used to lend it elasticity. Waste oils contaminated with PCBs were commonly sprayed on unpaved roads until a few years ago to help keep dust down.

Estimates of the total amount of PCBs manufactured and distributed in products like these in this country over the last half century run into the hundreds of millions of kilograms. The EPA estimates that some 335 million kilograms are still in use, and some 10 million kilograms are in storage, waiting for disposal. No one knows how much PCB has escaped into the ambient environment—estimates range from 45 million kilograms to as much as five times that. Most of what is classified as being loose in the environment is buried in old landfills, but that is not much reassurance, considering what we're learning about how poorly most landfills were constructed years ago and how susceptible groundwater is to irreversible contamination by persistent synthetic industrial chemicals. Furthermore, we probably don't even know where most of these landfill sites are anymore. It seems certain that more PCBs will be escaping into the biosphere in years to come as old electrical equipment long buried and forgotten in landfills across the nation slowly corrodes and begins to leak and as trucks carrying shipments of waste PCBs to safe disposal sites overturn and spill their cargoes.

PCBs are already so widely dispersed in the environment that they are routinely found in air and water samples all over the world, including five meters under the ice at the South Pole.

PCBs are toxic. We know that acute exposure to them can cause swollen eyelids, a waxy secretion from the tear glands, darkened skin and nails, severe headache, sore throat and cough, ulcers, and a variety of other gastric disorders, neurological problems, fatigue, impotence, jaundice, and menstrual irregularities. PCB exposure also causes a distinctive disorder called chloracne. Chloracne, which looks something like a bad case of teen-age acne, attacks the face, neck, and shoulders. Although it manifests itself in the skin, it is a systemic response of the whole body to exposure to any number of different chlorinated compounds. It is very disfiguring and can be extremely hard to get rid of, lasting for the victim's entire lifetime in some cases.

Chronic exposure to PCBs in lab animals has been shown to cause liver damage, gastric ulcers, and a whole gamut of reproductive disorders, including depressed birthrates, higher spontaneous abortion rates, and increases in birth defects, including some of the same kinds of thinning of eggshells that made DDT so hazardous to a number of bird species. PCBs are also mutagenic and carcinogenic in lab animals.

PCBs bioaccumulate in the fat of all living organisms. They also move up the food chain as one species consumes another, becoming ever more concentrated as they go, in the process known as biomagnification. A peculiar characteristic of PCBs in water makes them especially likely to transfer into organic matter in the water, such as fish, and creates an especially high risk of PCB contamination for any creature that eats much fish taken from areas of high PCB pollution.

On the other hand there is evidence from a recent study by James Allen and two associates that health disorders caused by ingestion of relatively low concentrations of PCBs are reversible, at least in adult monkeys, once PCBs have been removed from the diet.[12]

Allen and his coworkers fed sixteen female rhesus monkeys a diet containing 2.5 and 5 ppm PCBs for eighteen months, during which time the monkeys were bred and bore six infants. Mothers

Photo credit: Sea Grant Association

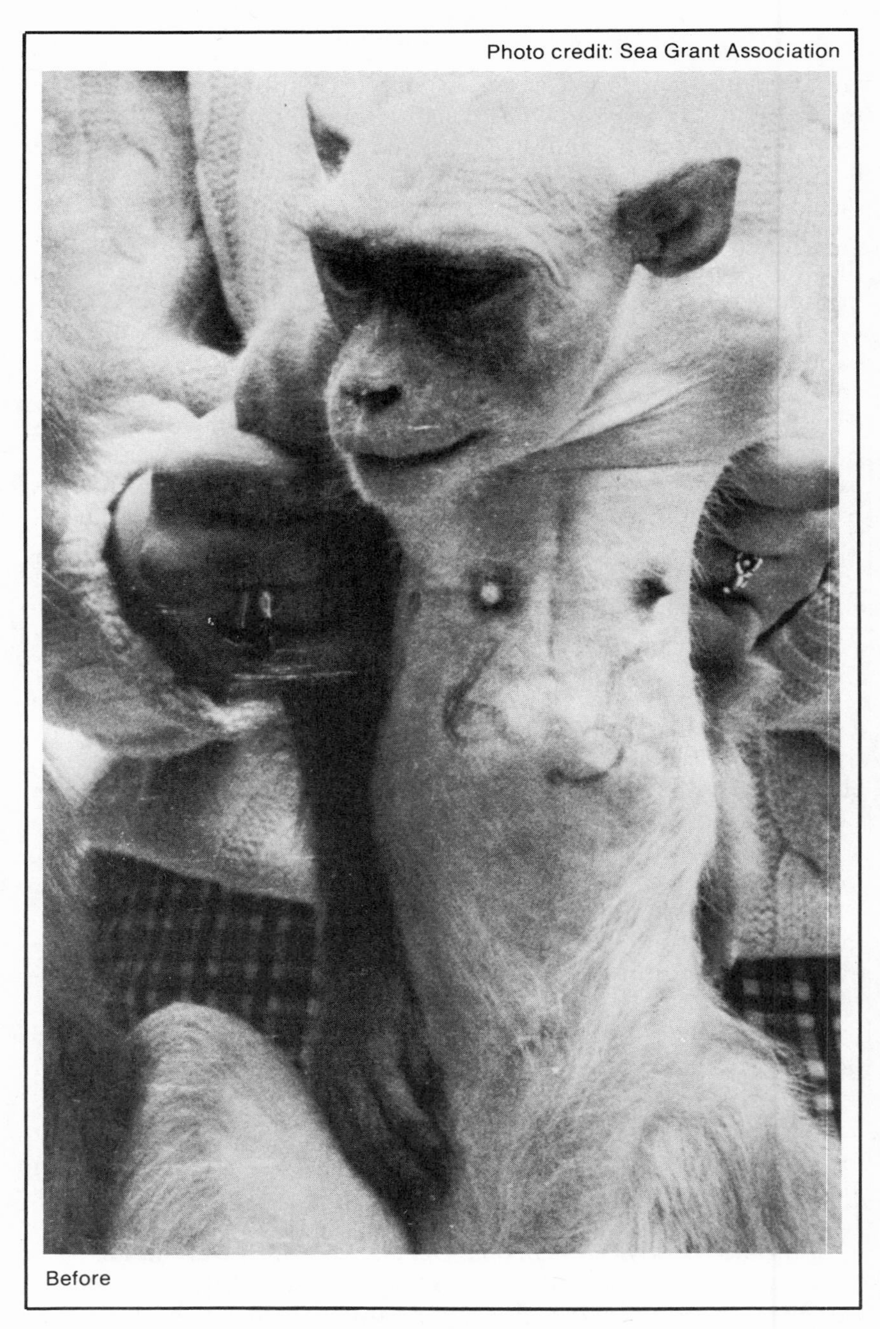

Before

Figure 10

Photo credit: Sea Grant Association

After

Figure 11

and infants alike showed strong signs of PCB poisoning, including loss of hair, swollen eyelids, and weight loss in the adults, and chloracne and skin darkening in their offspring. Three of the infants died within a year from PCB-related effects.

Once they were taken off the PCB diet, the adult primates improved steadily, until by one year afterwards, they were essentially in normal good health again. The photographs on pages 74–75 vividly illustrate the differences between poisoned and recovered monkeys.

YUSHO

Though we have known since the 1930s of the acute effects that workers exposed to PCBs could suffer, it took a mass public poisoning incident in Japan in 1968 to arouse widespread concern about the hazards of PCB ingestion. On a February weekend in 1968 a heat exchanger in a canning factory developed pin-sized holes that allowed PCBs to leak out into a batch of rice oil. On Monday the oil was canned, shipped as usual, and sold in stores throughout western Japan before the leak was found. Subsequent analyses of samples of the oil showed it to contain as much as 2,000 to 3,000 ppm of PCBs. By comparison the maximum concentration level allowed in a variety of foods today by the FDA ranges from 0.2 ppm in baby food to 5 ppm in fish and shellfish. However, small levels of other very highly toxic substances known as dibenzofurans were also found in the contaminated rice oil, clouding the issue of what really caused the subsequent outbreak of health disorders.

After a month had passed the symptoms of what eventually came to be known as "Yusho" (meaning "oil disease') appeared in the first of what are to date some sixteen hundred victims in Japan. Their symptoms included swollen eyelids, waxy secretions from the eyes, fatigue, headache, cough, abdominal complaints, menstrual disorders, peripheral nervous system problems, and chloracne. Investigators estimated that ingestion of as little as 200 to 500 milligrams of PCBs caused signs of the disease to appear in these individuals, though the average amount consumed was probably more like 800 to 2,000 milligrams.

Over time, signs of chronic PCB poisoning, including birth defects and cancer, began to appear in Yusho victims as well. Babies born to Yusho mothers after the incident were abnormally small, exhibited the characteristic waxy discharge from the eyes, and were colored a deep brown or grayish color. In some the bones of the skull took an abnormally long time to close after birth. Premature eruption of teeth was also noted in two Yusho infants. Tests revealed PCB levels ranging from 0.1 to 1.8 mg/kg in their tissues. Even some born before the poisoning took place were afterwards affected by ingesting PCBs in their mothers' milk.

Today, nearly a third of Yusho victims have reported varying disorders of the peripheral nervous system. Typical complaints include decreased sensitivity to pain, touch, and temperature. The question of just how carcinogenic PCBs are in humans remains unanswered to date. As of 1977, 35 percent of the deaths occurring among Yusho victims were due to cancer. Though this is substantially higher than the 21 percent cancer death rate among the total population in the area of Japan where Yusho struck, it is too soon to tell what part, if any, of this elevated cancer mortality figure among Yusho victims is due to PCB ingestion.

Surveys of the Japanese environment prompted by the Yusho episode turned up surprisingly high levels of contamination in some areas. A 1972 survey of the staple of the Japanese diet that is also most highly susceptible to PCB contamination, fish, showed that 16 percent of Japan's marine fish and 18 percent of its freshwater fish had at least 1 ppm PCBs in the tissues of their edible portions. Some fish, taken from highly polluted waters, had as much as 3 ppm PCB contamination. Today, fish caught in coastal waters around Japan are banned from market if they contain more than 3 ppm PCBs.

In June 1972 alarm over PCB contamination in Japan led to the banning of their domestic manufacture and, three months later, of their import. In 1973 Japanese industries still using equipment containing PCBs were advised to begin conducting regular health screening exams among their employees to detect early PCB poisoning.

In the United States the first steps toward regulating the manufacture and distribution of PCBs were taken, commendably enough, by the major producer of PCBs in this country, the Monsanto Company. As early as 1971 Monsanto was limiting the sale of its PCB products to firms that planned to use them only in equipment where they would be enclosed in such a way as to prevent their accidental release into the ambient environment. The following year Monsanto began marketing a PCB product that was significantly more biodegradable than its predecessors, and it restricted PCB sales to companies that serviced or manufactured transformers and capacitors. It also voluntarily contacted businesses that had previously used

Monsanto PCBs in heat exchangers or hydraulic equipment and advised them to drain the PCBs out and begin using non-PCB substitutes instead.

By 1973 the first federal regulations on PCBs were set by the FDA, which adopted temporary maximum contaminant levels for various foods. Before long a number of incidents involving contamination of animal feed and fishmeal from leaking PCB equipment led the FDA to ban the use of PCBs in any equipment used in food processing and animal feedlots or feed manufacturing plants. As late as 1977 the FDA was still finding PCBs in food samples from all across the country at levels high enough to trigger orders for the destruction of thousands of kilograms of chickens, turkeys, and eggs.

THE TOXIC SUBSTANCES CONTROL ACT

By early 1976 concern about man-made toxic compounds like PCBs was so widespread that both houses of Congress had proposed bills aimed at forcing manufacturers of these substances to bear the burden of extensively testing them before marketing them to prove that they would not be harmful to health and the environment. By late 1976 both bills had been combined into the Toxic Substances Control Act (TOSCA), which assigns the EPA broad regulatory powers over all new chemicals except for foods or food additives and a few other substances, such as nuclear materials, which are regulated by other agencies. TOSCA basically requires that anyone planning to introduce a new chemical into general use, or divert an old one to a new use, notify the EPA of their intention and prove to the EPA's satisfaction through extensive toxicological testing that the substance is safe. Furthermore, anyone who is allowed by the EPA to manufacture or use a potentially harmful chemical substance must keep extensive records of the amounts produced or used, how many employees are exposed to it, and for how long, how it is disposed of, and any complaints or allegations from any and every source at all about its impact on health or the environment. Violations of the act carry penalties of up to twenty-five thousand dollars per day in fines and up to one year in prison.

One hazardous substance, and only one, is mentioned by name in TOSCA—PCBs. On March 26, 1976, PCBs earned the dubious distinction of being singled out by Wisconsin senator Gaylord Nelson, who called in TOSCA section 6(e) for a ban on their production, processing, and distribution.

On October 11, 1976, TOSCA became Public Law 94-469. (Within weeks Monsanto had expeditiously cleared its shelves of PCB products.) But the actual ban took some time to go into effect, and there are serious questions today whether it ever really occurred at all. The EPA began promulgating regulations limiting PCB production and use within months after TOSCA was enacted. By February 1978 it had established requirements for labeling PCB products and for disposing of PCB wastes. Final regulations prohibiting any further manufacture, processing, or distribution of PCBs were not issued until July 1, 1979.

Unfortunately, TOSCA section 6(e) contained a very large loophole: Senator Nelson had included an exemption for PCBs used in "totally enclosed" systems, such as transformers, explaining that the way to judge whether a PCB product or system was "totally enclosed" was to determine whether it "will ensure that any exposure of human beings or the environment to a polychlorinated biphenyl will be insignificant."

The EPA chose to interpret the "totally enclosed" criterion quite liberally—so liberally, in fact, that it in effect exempted all of the hundreds of thousands of operating capacitors and transformers in the country, in which nearly all of the PCBs still in use are con-

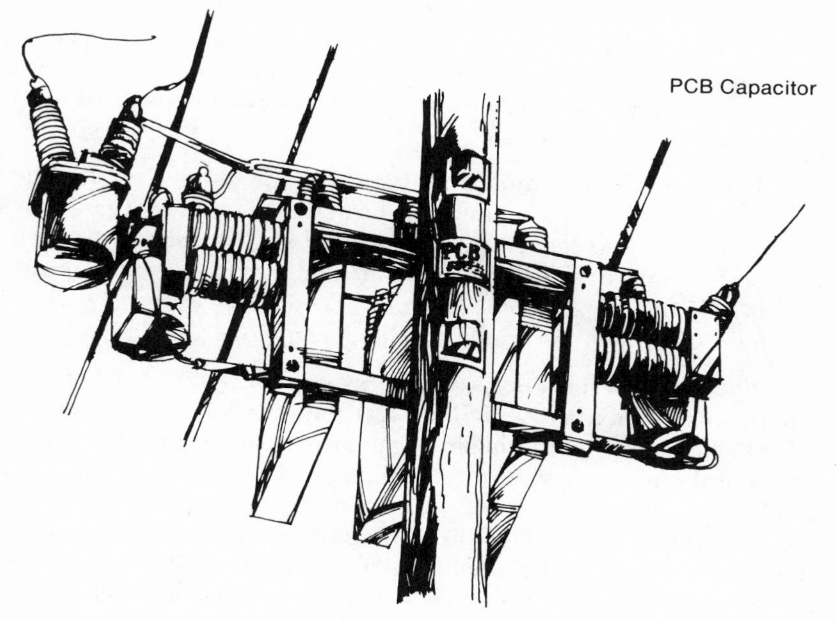

Figure 12

tained. It also went one step further and exempted *any* mixture, enclosed or not, containing less than 50 ppm PCBs.

In some ways this was a sensible thing to do. After all, in 1978 there were some 100 million capacitors in use in this country, virtually all of which contained PCBs. There are fewer transformers, but each one holds quite a bit more PCB coolant than the average capacitor, anywhere from one hundred to three hundred liters, and transformers have a much longer life span, up to thirty years, as compared to ten or so for the average capacitor. (Experts estimate that about two-thirds of all the PCBs still in use are in capacitors and most of the remaining third is in transformers.) Enforcing a PCB regulation that covered all these pieces of equipment would have clearly exceeded the EPA's limited enforcement resources.

Unfortunately, neither transformers nor capacitors are, by any stretch of the imagination, "totally enclosed." Both tend to leak, and, when used on utility company power lines in some parts of the country, to explode. Since power poles are commonly found in close proximity to schools, playgrounds, parks, hospitals, homes, and so on, human exposure in such cases is hardly "insignificant." The question of exactly how harmful is the exposure that is caused by, say, a capacitor that blows up and sprays PCB oils over nearby cars, lawns, and passersby remains to be scientifically resolved, but since PCBs are carcinogenic, the most prudent assumption would be that *any* exposure is significant. And, ironically enough, this is precisely the position that the EPA itself has taken on the issue of "safe" thresholds of PCB exposure:

> The available data indicate [that] PCB may cause several adverse effects in humans, mammals, birds, and aquatic organisms at extremely low concentrations. Therefore, for all practical purposes, exposures of humans and other animals to any level of PCBs should be deemed significant. This is especially true in light of the demonstrated carcinogenicity of PCBs.[13]

The EPA's second decision, the exemption from TOSCA regulations of any mixture containing less than 50 ppm PCBs, seems to have been equally questionable. Again, the basis for this decision was a practical one: there are literally millions of kilograms of weak PCB fluids in an enormous variety of pre-1979 consumer goods, including everything from color TVs to microwave ovens to ham radios to fluorescent lights, all containing small amounts of PCBs at concentrations less than 50 ppm. It would be an impossible task to even inventory them, let alone remove their PCB fluids and replace them with something safer. However, as these appliances wear out they will be thrown away, and, depending on how they are disposed of, in time their PCBs may leach out into the ambient biosphere.

In late 1979 concern about these loopholes in the EPA's PCB regulations and the continued problem of fresh PCB contamination of the environment moved the Environmental Defense Fund to petition the U.S. Court of Appeals for the District of Columbia to force the EPA to revise these regulations. After hearing the issue on June 6, 1980, the court concurred with the EDF. The "totally enclosed" exemption and the 50 ppm cutoff were ineffective, it said, and it ordered them set aside.*

The court also sharply rebuked the EPA for not more vigorously implementing Congress' desire to bring PCBs under stringent control:

> Human beings have finally come to recognize that they must eliminate or control life threatening chemicals, such as PCBs, if the miracle of life is to continue and if earth is to remain a living planet. This is precisely what Congress sought to do when it enacted section 6(e) of the Toxic Substances Control Act. Yet, we find that forty-six months after the effective date of an act designed to either totally ban or closely control the use of PCBs, 99% of the PCBs that were in use when the Act was passed are still in use in the United States. With information such as this in hand, timid souls have good reason to question the prospects for our continued survival, and cynics have just cause to sneer at the effectiveness of governmental regulation.
>
> The EPA regulations can hardly be viewed as a bold step forward in the battle against life threatening chemicals. There is no substantial evidence in the record to support certain of the EPA regulatory enactments, and portions of the regulations are plainly contrary to law. Thus, the effort by the EPA has, in certain respects, fallen far short of the mark set by the Congressional mandate found in section 6(e) of the Toxic Substances Control Act.[14]

In short, despite the "ban" on PCBs, they still constitute a significant pollutant of the environment. And this is a pollutant that

*On February 12, 1981, in response to a motion filed by the EPA and other parties, the same court granted the EPA a period of eighteen months in which to study the totally enclosed concept further and develop more appropriate rules to regulate these types of PCB equipment. The order setting aside the 50 ppm threshold was also stayed for an indefinite period of time while the court considers an EPA proposal for dealing with this issue.

does not seem to be abating. In fact, each year seems to bring us new examples of the pervasiveness of PCBs in the environment. For instance, a study of 1,957 lactating women in Michigan conducted in 1977–1978 found that *every one* of them had detectable levels of PCBs in her milk, ranging from trace amounts to over 5 ppm.[15] Half of the women tested had levels higher than the 1.5 ppm maximum allowed by the FDA in cow's milk, raising the sobering question of whether it was safe for these women to continue to breast-feed their own babies. The researcher conducting the study, Thomas Wickizer, recommended that new mothers with breast-milk PCB levels above 2.3 ppm halve the length of time they would otherwise nurse their babies, weaning them to cow's milk sooner. They also recommended that pregnant women avoid eating fish taken from waters polluted by PCBs, and that they avoid sudden weight loss, which can release large amounts of chemicals stored in fat into the bloodstream and thus into the placenta and the developing fetus.

PCBs may constitute a hazard in the ambient environment, but those who have the most to fear from this toxin are those who have to work with it regularly, maintaining and repairing the PCB equipment still in service. And since most of this equipment can be found at the top of utility poles, it is the workers who climb these poles for a living who face perhaps the greatest risk of exposure to the PCBs they contain.

Allen Simontacchi, a thirty-four-year-old utility lineman, has worked for Pacific Gas & Electric (PG&E) for the last twelve years. He lives in Sonoma, a small town about fifty miles north of San Francisco, with his fiancee, Chambier Bechtel, and their three children.

On January 16, 1980, at three forty-five P.M., the PG&E office in Sonoma received a phone call reporting an explosion on a utility pole at the corner of Napa and East Third streets. When PG&E's troubleshooter arrived at the scene minutes later, he found that the middle capacitor in a bank of three near the top of the pole had ruptured, spewing PCB-contaminated insulating oil all over the pole, the ground and sidewalk beneath it, on bushes and trees, and on a nearby rock wall and metal handrail. At four thirty, three PG&E workers were dispatched to the scene to clean it up. One of them was Allen Simontacchi.

> A few months before that spill an article had appeared in our union's monthly newsletter, describing PCBs—how they're carcinogenic, how they can be brought home to the worker's

family, and how they're used so widely in the electrical industry, in transformers and capacitors especially. This was the first time I'd seen anything on PCBs, and it stuck in my mind, naturally. So when the "cap" unit blew the following January, and I was sent out on the cleanup crew, I was concerned. The first thing I did was ask my supervisor whether we were supposed to notify the EPA. About an hour later he called back on the radio and said he'd checked with his superiors and they'd said it wasn't necessary. None of us on that crew really knew what we were dealing with or what to do about it. We hadn't had any training in cleaning up PCBs, and all we knew about their hazards was what we read about in the union paper.

We did what we could to clean up the spill. We spread Drizit [an absorbent powder] all over, cleaned it up with a power cleaner, then did all that a second time. We dug up some of the dirt directly under the pole, spread some Drizit down there, and left it there to soak up PCBs. Then, about eleven o'clock, we put up the barricades and the barricade tape and left.

But the leaking capacitor was still up there, dripping PCBs, and that bothered me. That night it rained, and since that pole is in a wooded area, right over a creek, I was sure that PCBs were being washed down into the creek that night. In fact I know that when the company sent a crew out the next morning to replace the ruptured capacitor, the pole was covered with oil when they got there. One of the men accidentally brushed against an oily tree branch and a few days later a rash broke out on his face and hands. None of them had ever been warned about PCBs by their supervisors, and none of them had ever been trained in how to clean them up safely either.

I was so concerned about this that I called the EPA myself the next day. Doing that scared me—I thought if the company found out about it I would lose my job. Then a couple days later I called the Fish and Game Department because I was still worried about the creek. They told me they would send out a field unit to check on the spill, but they never did. There was no follow-up on either of those calls.

A week went by and still nothing had happened. I didn't think the site had really been cleaned up enough. There was still oil all over the trees and bushes out there. So, the next weekend I was sitting around in my apartment, drinking wine with Chambier, and I told her about the whole thing. Pretty soon, she became curious and we went out to the site and to another one where the same thing had happened. Well, that's when things started to happen. As soon as she saw what was going on she got on the phone to EPA offices in Washington and she didn't stop until she had been referred to a local number. She called it and got ahold of someone and told him that if we didn't get some action within forty-eight hours she was going to the press.

The next day we started getting calls from local EPA people. Three weeks later an EPA field investigator came out and took samples at the Napa and East Third site. When they got the results they found that the site was still highly contaminated—over 80,000 ppm PCBs. They found the same thing at some other sites around here where capacitors had ruptured and had supposedly been cleaned up. All of them had illegally high PCB residues, and one of them was a place where children play. Another was a school bus stop. PG&E later paid some twelve thousand dollars in EPA fines for not doing a better cleanup job on these spills.*

On January 23, 1980, I filed a grievance through my union, Local 1245 of the International Brotherhood of Electrical Workers, claiming that we were not being trained adequately to clean up PCB spills. Before I filed it I went down with my union rep to discuss it with the PG&E division personnel manager, and to ask him some questions I had about handling PCBs. The answers he gave me were revealing. For instance, I asked if the company notified telephone workers when there was a PCB spill on a pole. He said, "No we don't because we're supposed to clean them up." But the answer that really bothered me was when I asked him what happens if a child is playing under one of these spills and gets saturated in PCBs. He said, "We just wash them off and pay them off." After that I went back and added a paragraph to the grievance calling for a phase-out of all PCB capacitors by PG&E.†

The EPA and the public put so much pressure on PG&E

*Chambier Bechtel later founded the PCB Project, a pioneering organization that offers information and support services to workers in the San Francisco area who are faced with toxic exposure problems. The project has done much to publicize the PCB issue in Northern California.

†On August 5, 1980, PG&E announced plans to phase out the one hundred twenty thousand capacitors in its service area containing PCBs over the next three years, at an estimated cost of some $50 million.

about these spills that it hadn't cleaned up safely that it had to go back to the Napa and East Third site five times before it really got clean. I was sent out on three of those times. After I came home the third time, the more I thought about how we kept having to go back there without adequate training or supervision, how we were working with this toxic stuff but we really didn't know enough about it or how to handle it, the more I realized I couldn't take it anymore. The fourth time they told me to go, I refused. So, I was suspended. The suspension lasted for two days and through my grievance procedure I got one day's pay back. Right after I refused to go, they asked another lineman to go, the one who had gotten the rash on his face, and he refused, too. After that they just quit trying; they didn't even bother to ask any more linemen to go. They sent a driver and a subforeman and that was it. They knew the rest of the men would refuse, too.

One of the main things bothering me was if I was going to have to work around PCBs then I at least wanted to have adequate protective clothing to wear, and that we didn't have. The protective clothing we were issued was so flimsy we called them pampers. They were just very thin paper coveralls, which tore very easily. Since everyone wanted the big sizes, half the time all they had were sizes that were too small. Those tore readily, since lineman are big men. Even when they're the right size, they don't completely protect your body—your neck and head are exposed. There weren't any respirators available at that time, and even if we'd had them, there wasn't anyone who could tell me when we needed them and when we didn't. All we had were the pampers, plastic boots, and rubber gloves. Just recently I found out the pampers are, literally, only good for about ten seconds of protection, if they aren't broken. They have what industrial hygienists call a ten-second "breakthrough time." After that, if PCB is getting on your skin, your body is absorbing it. And if it's getting on your clothes, you better take them off before you go home or you'll be exposing your family, too. We hear that someone is working on a suit that will have a one-hour breakthrough time, but that's still not good enough. What is the company going to do if the job takes five men a total of six hours to do? Is it going to issue us the thirty protective suits that it would take to do the job safely? I feel strongly that if supervisors are ignorant of toxins they should be educated. If they don't want to be educated, they should be removed from their positions. And if they *do* know about toxins and aren't informing their workers, they should be brought up on criminal charges.

From March 1980 to May 1981 I was sent out at least five more times to help take down PCB-contaminated cap banks. And the company still isn't taking this problem seriously.

What I resent is that I'm harassed for questioning these things, for being concerned about my health. And it's not just me, just one person—there must be five thousand of us in the electrical industry in the Bay Area alone who are exposed to PCBs.

But it's tough to reach a lot of these guys. I want to grab some of them and say, "Wake up." But when you've got a mortgage and a family, the first thing you do is try to deny it's that bad. You think, the company wouldn't do that to me, not after all those years. Or there are guys who just flat out deny that PCBs are dangerous. They say, "Oh I've been working up to my elbows in that stuff for years and it hasn't bothered me." Well, maybe not yet, I always think.

The way it is now, if I refuse to go out on PCB assignments I'll lose my job. It's a matter of having a job or not having a job. And with three children, that is something that works on your mind considerably, believe me.

I've always been very active in my union—I've been a shop steward for seven years—but I don't feel we've gotten much cooperation from my local. I don't know why . . . it's sad. Health and safety on the job should be something you don't compromise. What good does it do to negotiate your health away in favor of better wages and fringe benefits if you're going to be dying of cancer at fifty-five? Maybe it's beginning to change. The guys I work with are getting scared and angry, the more they hear about PCBs and other toxins.

There are substitutes for PCBs, many of them. One called DOP (dioctyl phthalate) has been in use in Japan since 1974, and is now often used in new capacitors built in this country. So far, DOP seems like a clear improvement over PCBs. It is less expensive and even less flammable than PCBs, and has comparable insulating properties. What little information we have about its toxicity indicates that it is far less hazardous than PCBs. Other proposed PCB substitutes include silicon compounds, synthetic hydrocarbons, and even fluorocarbons, whose use in aerosol propellants was banned in 1978, because of fears of ozone damage. From an environmental health perspective, the most attractive substitute is probably mineral oil, the least toxic and most biodegradable of any of these alternatives.

However, finding acceptable substitutes for the PCBs still in use is only half of the problem we face. We still have to dispose of the PCBs we drain from our capacitors and transformers, and until recently that was not an easy thing to do. Very weak concentrations of PCBs can sometimes be buried in secure landfills, but most others cannot, because of the possibility of adding to the significant environmental PCB burden we've already created. Incineration is the logical disposal method in these cases, but until recently it posed special problems of its own: under all but the most rigorously controlled conditions at exceptionally high temperatures, combustion of PCBs can cause the formation of two other compounds—dioxins and dibenzofurans—that are far more toxic than PCBs are. Consequently, until early 1981, there was not a single commercial incinerator in the country that the EPA felt was equipped to handle this task safely.

But on January 21, 1981, an incinerator located at Deer Park, Texas, operated by Rollins Environmental Services, Inc., was approved by the EPA for burning PCBs, and a second one, located at El Dorado, Arkansas, and operated by Energy Systems Company (ENSCO) was approved shortly afterwards.

Both these incinerators do release tiny amounts of dioxins and dibenzofurans. However, careful sampling of the stack gases during test burns and calculations of the amount of dioxins and dibenzofurans emitted convinced the EPA that in the *worst possible case,* exposure to these emissions would mean a potential increase of less than 1 additional cancer case per 50,000 exposed people in the vicinity of the Rollins facility and 1 additional cancer case per 2.4 million people around the ENSCO site, risks that the agency considers justified by the benefit of having a way to dispose of PCBs.[16]

There may be another, even safer way of getting rid of some types of PCB compounds. The Sunohio Corporation of Canton, Ohio, a joint venture of the Sun Oil Company of Philadelphia and the Ohio Transformer Company, won approval from the EPA in late May 1981 to begin operating a unique, patented system for destroying PCBs that involves no combustion at all. The details of the system are secret, but in general it involves the use of a chemical reagent known as PCB-X that separates the chlorine atoms from the biphenyls, turning them into harmless chloride salts, and producing a tar-like residue that can be disposed of in conventional landfills. The process does have limitations: It can handle any liquid PCBs that are not contaminated with water, but it cannot be used to destroy contaminated rags, clothing, soaked earth, and so forth. The entire Sunohio system is housed in a large trailer that can be hauled from place to place as needed. Its main purpose, according to Sunohio, will be to decontaminate transformer oil, which can be reused once it has been cycled through the system. As well as being safer, the cost of the Sunohio process will probably be significantly lower than incineration.

It is only in the last decade that a concerted national effort to protect the men and women who work in industrial settings from toxic substances has begun to take shape in this country. Before 1970, worker health and safety was left to the individual states to safeguard, and in many of them next to nothing was done about it. In fact, in some parts of the country, the very concept of occupational disease was met with outright scorn. *America's Textile Reporter* editorialized on June 10, 1969:

> We are particularly intrigued by the term "Byssinosis," a thing thought up by venal doctors who attended last year's ILO [International Labor Organization] meetings in Africa where inferior races are bound to be afflicted by new diseases more superior people defeated years ago.
>
> As a matter of fact, we referred to the "Cotton Fever" earlier, when we pointed out that a good chaw of B.L. Dark [a brand of chewing tobacco] would take care of it, or some snuff.

On December 29, 1970, the Occupational Safety and Health Act was signed into law. It took effect 120 days later, on April 29, 1971. To carry out its mission of assuring worker safety and health, the act established three complementary federal agencies. The first, the enforcement arm, called the Occupational Safety and Health Administration (OSHA), became a part of the Department of Labor. The second, the research branch, known as the National Institute of Occupational Safety and Health (NIOSH) was incorporated into the Center for Disease Control, which is part of what has recently become the Department of Health and Human Services. A third, the Occupational Safety and Health Review Commission, an arbitration body whose members are appointed by the President, hears disputes arising from enforcement of the OSHA act.

In general, NIOSH identifies and researches occupational health problems in order to find ways to prevent or mitigate them. Its studies usually result in recommendations that workplace exposures to a specific substance be kept below a certain numerical level. These recommendations are then handed to OSHA, which has the power to make these exposure recommendations legal standards and to see that they are upheld by employers. Normally, the process of setting a standard is a long and cumbersome one, but the secretary of labor does have the power to set a temporary emergency standard, which takes effect almost immediately, if the circumstances are sufficiently hazardous. Public employees are not covered by OSHA, but most other workers, a total of about 60 million Americans, are.

Not surprisingly, OSHA has often met with a less than enthusiastic response on the part of business. Many of its regulations have been appealed and some have been overturned. Its most appreciative response has come from labor.

Nonetheless, in the decade following passage of the act, an unprecedented wave of concern about occupational health and safety issues swept across the country. As a consequence unions, government, and even industry made great strides in evaluating and controlling workplace hazards. In all these sectors there was a sharp increase in the number of personnel assigned to deal with these issues.

The OSHA/NIOSH record to date is mixed. Support for their activities varies greatly from one presidential administration to the next. Annual efforts by Congress to repeal or significantly amend the act in ways that would weaken it have become routine. Top OSHA/NIOSH administrators are regularly replaced, and cooperation between the two agencies has been uneven. The costs of researching workplace hazards has increased tenfold while, according to some critics, the quality of the research performed has steadily declined.

Though clear-cut advances have been made since 1970 in improving workplace health and safety conditions, significant problems remain. Employers often cooperate only grudgingly, at best, with regulations. Workers themselves fail to follow stringently health and safety practices devised for their protection. Relatively few of the thousands of compounds routinely used in industrial settings today have yet been adequately tested for toxicity, particularly for chronic effects. In most states the Workers' Compensation system is still primarily aimed at offsetting the costs of traumatic injury rather than long-term, cumulative disease. Finally, the system established under the 1976 Toxic Substances Control Act, described earlier in this chapter, for testing new compounds for toxicity before use still has serious shortcomings.

With the Reagan administration have come sharp cuts in funding and staffing levels for OSHA and NIOSH programs. What impact these reductions and the attendant shift in official attitudes toward occupational health and safety issues will have on the private sector remains to be seen.

4

HAZARDS IN THE AIR

On October 9, 1980, during a campaign stop at a dying Ohio steel mill, presidential candidate Ronald Reagan was quoted as saying that "air pollution has been substantially controlled."[1] Two days later Reagan's flight home to Burbank Airport in southern California had to be diverted to Los Angeles International because the smog around Burbank was so thick that visibility was nil.[2]

Unfortunately, air pollution has *not* been controlled in America today, and there is good reason to fear, due to actions taken by the Reagan administration itself, that the significant progress we've made toward controlling it over the last decade may be lost in the next few years.

What *is* true is that we know more about air pollution today than ever before, and we're learning new things about the very complex chemistry of the atmosphere each year. We know what the worst pollutants are, where they come from, and how to reduce—but not completely eliminate—most but not all of them. We also know that polluted air does cause death, disease, and billions of dollars in property damage, cleaning costs, and lost crops each year. We also know that, as long as we continue to store, handle, and burn significant amounts of fossil fuels, the need to control the pollutants they and many other byproducts of industrial technology produce will always be with us.

AIR

Our atmosphere, a thin (thirty kilometers or so in depth) membrane of gases enveloping the earth, is composed mainly of nitrogen and oxygen, plus small percentages of water vapor and carbon dioxide, and fractional amounts of a host of other gaseous, liquid, and particulate substances. It is essential to our survival, not only because we must breathe it to live, but also because it shields us against the hazards of outer space. Without the atmosphere around the earth, the sun's heat would fry whichever side was exposed to it, leaving the other side in freezing cold. The ozone layer in the upper half of the atmosphere (known as the stratosphere; the lower half is the troposphere) protects us from excessive exposure to the sun's ultraviolet rays: without it we'd all contract lethal cases of skin cancer. The atmosphere also protects us from the showers of small meteorites and other galactic debris that constantly bombard the earth, by creating friction that burns them up before they can reach the earth's surface. Last but hardly least, the atmosphere screens out much of the cosmic radiation the earth is constantly bathed in, keeping our exposures to low, tolerable levels.

The nitrogen that we inhale is a fairly stable gas; it does not react readily with other substances under most conditions. Oxygen, on the other hand, is highly reactive, combining easily with many organic and inorganic compounds in a process known as oxidation. And it is a good thing for us that oxygen is so reactive: It is oxidation in our cells, taking place constantly at an incredible pace, that provides the basic energy needed to fuel the human metabolism.

Urban air, especially in industrialized societies, contains hundreds of other substances as well, many of which are less than beneficial to us. In the United States, air pollution control efforts focus on half a dozen broad pollutant categories, though there are hundreds of substances that have at one time or another been categorized as air pollutants, and new or better definitions of the old standbys appear each year. In this chapter we discuss five of the primary pollutants—carbon monoxide, sulfur oxides, nitrogen oxides, photochemical smog, and total suspended particulates—plus a few others.

On the average the concentrations of these pollutants in urban air is hundreds or even thousands of times their levels in pure air. Some of them dissipate in a few hours or days, but others may remain in the atmosphere for years, building up to higher and higher levels as time passes. Though the absolute amount of these substances is generally quite low, even in the most polluted areas, we are so

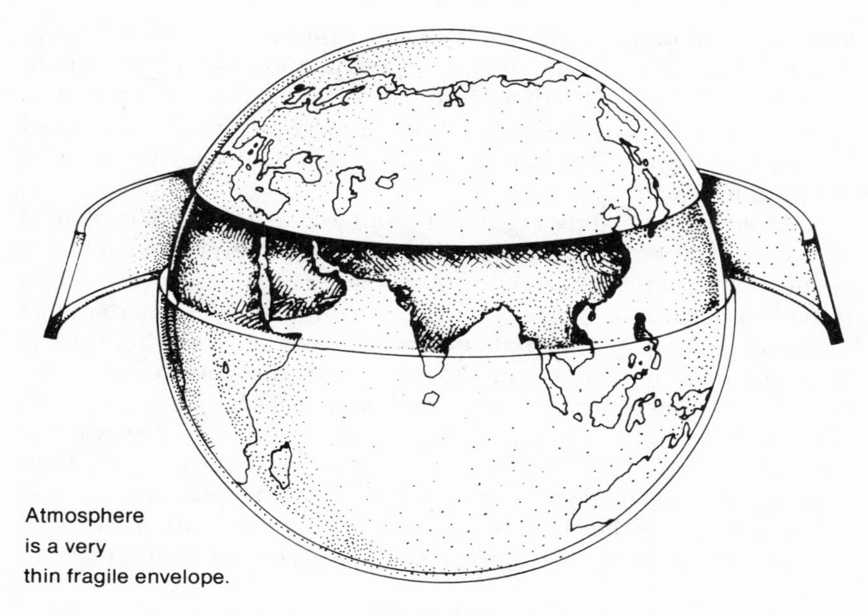

Figure 13

sensitive to the air we breathe that these very small concentrations may be hazardous. Ozone, for example, the major constituent of photochemical smog, probably never exceeds 0.5 ppm of even the dirtiest urban air, but will cause irritation and significant distress at much lower levels than that, as low as 0.15 to 0.20 ppm for asthmatics and other especially sensitive people. It is also important to remember that, while tests of the health effects of air pollutants are conducted with pure compounds, under real-life conditions there are hundreds of different substances in the air and many of them interact in highly synergistic ways. It has, as we explained in Chapter 2, long been known that the presence of respirable particles in air aggravates the effect of sulfur dioxide on the respiratory system, and more recent evidence suggests that sulfur dioxide enhances the irritating effect of ozone. But the synergistic interactions of air pollutants is probably the area we know least about.

The argument is often heard that nature herself is a much greater "polluter" of air than man is—and in a very simplistic, misleading way this is true. For every ton of pollution man puts into the air, nature adds a hundred more. Volcanoes, forest fires, decaying organic matter, pine trees, sea spray, and natural erosion of the minerals in rock and soil that then leach into waterways and are borne aloft by evaporation, only to fall back to earth later as the water vapor around them condenses into precipitation—all these sources of "pollution" far outweigh even the worst of our industrial polluters in terms of volume. Only in a few areas, such as the production of carbon dioxide and sulfur, does mankind appear to be approaching the output of nature.

However, what this argument overlooks is (a) the fact that in and around cities, where most of us now live and even more will live in the future, the proportion of man-made pollutants far outweighs those from natural sources, (b) that man-made pollutants are the only ones we can do much about, and in city after city where something has been done to control them, a clear, direct relationship between control measures and lower pollutant levels has been recorded, and (c) that nature has an abundance of highly effective ways of controlling her own pollutants—what she generates, she usually finds an efficient way to dispose of—and her natural pollution control mechanisms have been working well for hundreds of millions of years. Only industrial man seems to have the potential to unbalance them.

The most common natural depolluter of the air is precipitation—plain old rain, snow, and sleet. Pollutant gases that are soluble in water may combine with the water vapor in the atmosphere and then wash right out of the air with the next rainfall. Unfortunately, in the case of sulfur and nitrogen oxides, this may mean that the pollution problem is merely relocated rather than eliminated, since these two gases are the precursors of "acid rain," a growing problem in many surface waters and one that we discuss further in the next chapter. Precipitation is most effective in cleansing the air of particulates; conversely, a high level of particulate contamination in the atmosphere may expedite precipitation by providing small collection sites on which water vapor and other gases can condense to form droplets.

The wind is another effective natural depolluter, though here again the results of its action are often as much a matter of relocating pollutants as of eliminating them. But by distributing pollutants widely around the earth, the wind helps to diffuse them and thus keep them from building up to hazardous concentrations in any one place. Many chronic smog problems in urban areas would be greatly reduced if we could depend on strong winds to break up temperature inversions—those perverse stratifications of the atmosphere that result in a layer of warm air sitting atop a layer of cooler air and pinning it in one place, where it steadily absorbs the pollutants we pump into it.

Like the rain, the wind also helps remove air pollutants by literally running them into the ground—or buildings, foliage, people, and so on—where they stick. The grime on your windowsill is evidence of natural wind-removal at work.

Finally, there is gravity. The very tug of the earth's mass helps remove pollutants—especially particulates—from the air, causing them, as they bump into one another and adhere, slowly growing larger and larger, to settle gradually toward the ground.

CARBON MONOXIDE

Man's ability to pollute air is as old as his ability to start a fire. Carbon monoxide (CO) is produced by the combustion of any fuel containing carbon, and that covers just about all of them—gasoline, oil, natural gas, kerosene, fuel oil, diesel fuel, coal, charcoal, and wood. CO is the most common pollutant emitted by your car. In one well-known caropolis, Los Angeles, an estimated 9,000 metric tons of CO a day is pumped into the air by the automobile. Of the 180 million metric tons of all the major air pollutants America puts into its air each year, nearly half is carbon monoxide, and nearly two-thirds of that comes from motor vehicles. CO is ten times as common as any other major air pollutant.

CO is invisible, odorless, tasteless. It gives no warning signals. It is especially hazardous when vehicles powered by gasoline or other carbonaceous fuels are operated in a closed place. Each year hundreds of workers are sickened by CO poisoning when gasoline-powered forklifts are used in warehouses, factories, and ships' holds without due regard for the CO they produce. (This is why electric-powered equipment is usually used in closed spaces.) Gas appliances used in homes without adequate ventilation are also potentially hazardous, especially if their own venting systems aren't working correctly. Even outdoors, CO can become a hazard in heavy traffic, especially at high altitudes. Cigarette smoke is also an important source of carbon monoxide. Smokers generally have more than twice the level of CO in their bloodstream than nonsmokers do. Also, the newer "low-tar" cigarettes, whatever else their merits may be, generally give off more CO than older brands did.

Inhaled carbon monoxide is hazardous because it passes easily through the lungs into the blood, where it combines with hemoglobin, the substance that carries oxygen. The result of this insidious merger is a new compound called carboxyhemoglobin, which does a much poorer job of conveying oxygen. As a result the heart begins to work harder to get oxygen to the rest of the body, with less and less success as more and more oxygen is inhaled. All this can happen very quickly in an atmosphere polluted with CO. At concentrations of a few hundred ppm, CO poisoning causes a sense of giddiness, especially during exertion, and a throbbing head. Up to about 1,000 ppm, the symptoms progress to weakness, dizziness, nausea and vomiting, and irritability. At concentrations up to 1,500 ppm, CO poisoning can cause collapse, and from there on, coma, convulsions, and death. Even if a victim of CO poisoning recovers, there are often protracted aftereffects, including in some cases damage to the central nervous system leading to intermittent muscle spasm and even paralysis.

In the general population, among those who are not likely to be found working with gas-powered vehicles in closed spaces, the greatest risk of CO intoxication is to smokers with preexisting heart disease caught in heavy traffic. At high altitudes, in polluted urban areas, CO concentration under these conditions can exceed 100 ppm. While carbon monoxide has been known to *cause* heart attacks, it is more likely to increase the severity of heart attacks caused by other factors, turning an otherwise mild or moderate attack into a severe or even fatal one. This is one reason why good physicians relentlessly urge their heart patients to quit smoking. Even in otherwise healthy nonsmokers there is some evidence that CO can cause deterioration in general alertness and exercise ability.

SULFUR OXIDES

Next to carbon monoxide the second most abundant air pollutant is sulfur compounds, a class that includes, at its broadest, sulfur trioxide, sulfuric acid, a group of sulfur salts, or sulfates, and sulfur dioxide (SO_2), which is easily the most common sulfurous air pollutant.

Unlike clandestine pollutants that give no warning of their presence, sulfur compounds are notorious for their sharp, suffocating fumes and highly distinctive odors. If you've ever struck an old-fashioned wooden kitchen match, you've smelled sulfur.

Like carbon monoxide, sulfur pollution has a long history. Pliny the Elder, a victim of Vesuvius' sulfurous fumes in A.D. 79, is reputed to have thought that sulfur caused epilepsy. As Londoners turned from charcoal to coal at the dawn of the Renaissance, many commented in disgust on the infernal stink and smoke of their sulfurous new fuel. In the seventeenth century Bernardino Ramazzini, an Italian physician long considered the father of occupational medicine, devoted a section in his classic work *De Morbis Artificum* to diseases of sulfur workers. Sulfur, Ramazzini noted, had been used since the days of the Roman Empire to bleach and fumigate clothing, and those who came into contact with it were known for their "coughs, dyspnoea [shortness of breath], hoarseness, and sore eyes."[3]

Today, for every metric ton of sulfur air pollution that mankind pumps into the air, nature adds another seven or eight. But we are rapidly catching up: By the end of this century, if current trends continue, industrial man will have become the primary source of sulfur pollution in the Northern Hemisphere. By the end of this century we will be releasing some 270 million metric tons per year into the biosphere, about the same amount nature emits annually.[4] Nearly all of our current sulfur emission is caused by the combustion of coal, oil, or other petroleum derivatives. Of a current total of some 37 million metric tons of sulfur pollution produced by America each year, nearly 80 percent comes from power plants and factories burning fossil fuels, and most of that is SO_2.

Generally, SO_2 doesn't remain in the air very long before it is either removed by natural cleansing processes or reacts with other chemicals in the air to form new sulfur derivatives. When, for example, it adds a molecule of oxygen to become sulfur trioxide (SO_3), then combines with water vapor, the result is sulfuric acid (H_2SO_4), the same substance that is in your car battery.

Cities in heavily industrialized areas of the Northeast commonly experience SO_2 levels up to 1.7 ppm, occasionally rising as high as 3 ppm or so, the level at which the sharp, pungent fumes of SO_2 in the air first become noticeable. At higher levels, starting at about 6 ppm, acute exposures to SO_2 can cause a burning sensation in the eyes and the mucous membranes of the nose and throat, as well as coughing, nosebleed, and a choking sensation caused by the body's automatic attempt to shut off its respiratory passages to this noxious invasion. At very high levels sufficient irritation of the lungs by SO_2 can cause fluids to accumulate in them, a condition known as "pulmonary edema," one that can lead to serious interference with normal respiration. Chronic exposure to SO_2 has been associated with chronic sore throats, diminished sense of smell, and corrosion of tooth enamel. As we explained in Chapter 2, the presence of particulate matter of a respirable size in the air increases the hazards posed by SO_2. The gas is so soluble it normally reacts with the moisture of the upper respiratory passage and becomes sulfuric acid, which is what causes the stinging sensation of SO_2 exposure. But if a molecule of SO_2 has first adhered to a respirable particle (the process known as *ad*sorption) before it is inhaled, the chances of its penetrating much deeper into the lungs, riding on the insoluble particle, are greatly increased, and much more severe complications of conditions like bronchitis and emphysema may result. The adsorption of SO_2 on particles of zinc ammonium is thought to have been responsible for the deaths of twenty people in a famous air pollution incident that took place in Donora, Pennsylvania, in 1948. SO_2 adsorbed onto particles of coal dust was implicated, as we shall see below, in the deaths of some four thousand residents of London during a severe coal smog in 1952.

NITROGEN OXIDES

The nitrogen that makes up most of the gaseous mixture we must breathe to survive is hardly a threat under normal circumstances. But in some circumstances, such as when it is heated, nitrogen may react with other gases, especially oxygen, to form a class of compounds called nitrogen oxides, some of which are quite hazardous. The most common of these is nitrogen dioxide (NO_2).

As is true of most other common air pollutants, natural sources of nitrogen dioxide produce far, far more of it each year than man does—at least 10 metric tons for every one that we generate. However, in and around heavily polluted cities that ratio can shift dramatically. There is evidence that as much as one-third of the nitrogen oxides in Los Angeles' air comes from the combustion of fossil fuels. In rural areas nitrogen dioxide can form at toxic levels in silos containing freshly cut grasses and grains. Known as "silo-fillers' disease," NO_2 poisoning is one environmental health hazard that occurs as often in rural environments as in urban ones.

Nitrogen dioxide is a heavy reddish-brown gas that tends to sink to the ground in the absence of wind currents. Although there have been reports of overall nitrogen oxide concentrations at the 3 ppm level in Los Angeles, urban residents are usually not exposed to levels higher than 1 ppm in the ambient air.

Like SO_2, NO_2 is hazardous because it can combine with moisture in the respiratory tract to form an acid, nitric acid (H_2NO_4). However, NO_2 is a more insidious toxin than SO_2 because this process takes place much more slowly. As a result exposure to NO_2 is not as immediately irritating as exposure to a comparable dose of SO_2. The victim does not feel the strongly irritating burning sensation in the throat and eyes that heralds SO_2 exposure, and, misled by a false sense of security, tends to remain exposed longer than is safe.

In workplace settings NO_2 concentrations begin to become acutely hazardous at 50 to 60 ppm. At levels of 100 ppm, NO_2 exposure causes a change in the chemistry of the blood known as methemoglobinemia, a condition in which the hemoglobin loses some of its vital oxygen-carrying ability.

Some three to twenty-four hours after exposures at these levels begin, when the acid-formation process has reached harmful levels, an exposed individual starts to experience the most common first symptoms of NO_2 poisoning, a shortness of breath from pulmonary edema caused by burned lung tissues.

If the NO_2 concentration is low enough and the exposure time is short enough, a victim of NO_2 poisoning will recover in three to thirty hours after symptoms first appear, with no long-lasting ill effects. But in a few cases, less than 10 percent of all those that occur each year, apparent recovery is followed by a severe relapse, caused by actual destruction of small air passages in the lungs and characterized by chest pain, coughing, bloody sputum, chills, fever, nausea and vomiting, a bluish tinge in the lips or fingernails known as cyanosis (a sign of methemoglobinemia), low blood pressure, agitation, anxiety, and in severe cases, death from respiratory failure.

Nitrogen oxides in general also play a role in the formation of smog, discussed below, and there is speculation that exposure to them in this form may exacerbate certain chronic lung diseases and reduce the body's natural capacity for defense against bacterial and viral infection. In some parts of the country, predominantly the West, nitric acid rain is a greater problem than sulfuric acid rain, because there the proportion of cars to coal-fired power plants is higher than in the East.

PARTICULATES

Urban air pollution in industrialized countries is never purely gaseous—it always contains particles of liquids and solids as well.

Water vapor, sulfuric acid, and nitric acid are classic examples of liquid particulates. Solid particulates can include an infinite number of different kinds of soot, dust, ash, metals, and even certain compounds, such as sulfates and nitrates, that started out as gases. The scientific term for any conglomeration of particulates, whether solid, liquid, or a mixture of both, is "aerosol."

Particulates enter the atmosphere from a variety of natural and man-made sources—sandstorms, sea spray, fog, volcanic eruptions, forest fires, power plants, manufacturing plants, automobile exhausts, spray cans, and the burning of agricultural or forest land to clear it. Most of man's contribution to the particulate load in the atmosphere comes from the industrial combustion of fossil fuels, though this is also the area where control of certain kinds of particulate emission has been the most successful. Cigarette smoking is also a significant source of particulates, worse than even the most heavily polluted industrial atmosphere, at least for those who smoke.

Fortunately, as we explained in Chapter 2, the respiratory system is well equipped to defend the body against most particulates, which tend to be larger in size than gas molecules. It is more often the insoluble gases that pose a health hazard than the particulates in the air. There are some exceptions, of course, such as asbestos particles, coal dust in mines, cotton fibers in textile plants, and lead dust or fumes, which we discuss below, but for most of us particulates pose a health problem only as couriers of other, more toxic gases and liquids adsorbed on to these microscopic specks and thus carried much deeper into the lungs than they would otherwise be.

LEAD

One of the most common and most hazardous particulate pollutants is lead. Lead (Pb) is a dense, soft, grayish, slightly soluble metal which has found endless useful applications—pottery, batteries, radiators, radiation shielding, plumbing, paints, glass, printing—throughout history and which since the Industrial Revolution has become a pervasive contaminant throughout the biosphere.

Studies of the skeletal remains of ancient peoples routinely show that they contain much less lead than the bones of contemporary humans do. For example, a 1979 report in *The New England Journal of Medicine,* by Jonathan Ericson and two coworkers, concluded that lead levels in Peruvians living some 4,500 years ago were probably about 1/100 the level commonly found in the bones of modern industrial man. Furthermore, the authors of the study argue quite convincingly that "present-day man is subjected to exposures that elevate concentrations of lead in skeletons about 500-fold above natural levels."[5]

Only about 20 percent of our lead exposure is due to air pollution. The rest comes from lead in our food and water and, for workers in certain industries, the occupational environment. Lead added to gasoline to prevent knocking is the major source of lead particulates in the ambient atmosphere.

Lead is hazardous because it can undermine the body's ability to form hemoglobin. It can also damage the kidneys, the central nervous system, and the reproductive system, causing such problems as impotence, sterility, miscarriage, stillbirth, and birth defects. Moreover, lead can accumulate in the body, especially children's bodies, building up to higher and higher concentrations with additional exposures, rather than being naturally excreted over time.

The blood of the average American adult contains about 10 to 15 micrograms per decaliter (μg/dl) of lead, a level that most researchers believe has an insignificant effect on health. Although individual sensitivity to lead poisoning varies widely, and we still haven't adequately tested the effects of chronic, low-level exposure to lead, most specialists in this area feel that lead poisoning does not become a significant problem for adults until bloodstream concentrations exceed 50 μg/dl. Children are much more vulnerable to lead poisoning, however, and begin to be adversely affected at about 40 μg/dl. Since the human placenta does not block lead, fetuses are even more susceptible to the lead that invades their mothers' bodies.

Few samples of urban air contain lead levels above 10 $\mu g/m^3$, and the norm is far below that. At these levels lead air pollution will not add appreciably to the blood lead level of the average adult, even over long periods of exposure. In the workplace, however, lead levels as high as 300 to 900 $\mu g/m^3$ have been recorded, and workers in industries like lead smelting and battery manufacturing have been known to develop blood levels of 80 to 90 μg/dl. Even workers in less polluted settings, such as toll collectors, police officers, and highway crews, may average blood lead levels approaching 40 μg/dl.

The symptoms of toxic lead exposure can, depending on individual sensitivity, include fatigue, apathy, drowsiness, and decreased mental alertness, as well as insomnia, gastrointestinal pain that is sometimes mistaken for appendicitis, nausea, vomiting, constipation, and irritability. In children hyperactivity is a common symptom of toxic lead exposure. Severe cases of lead poisoning are marked by tremors and convulsions, especially in children, and can be fatal.

There is also strong evidence that low-level exposure to lead can cause intelligence deficiencies and behavioral disorders in children. A 1979 study by Dr. Herbert L. Needleman of 158 first- and second-graders related high blood levels (over 20 ppm in teeth—probably equal to about 25 to 45 μg/dl in the blood) with poor scores on standard intelligence tests and "teachers' reports of increased distractibility, increased prevalence of daydreaming, lack of persistence, inability to follow directions and lack of organization."[6]

The attempt to control lead emissions into the air by removing lead additives from gasoline and mandating air pollution control equipment that cannot tolerate leaded gas is a step in the right direction. Unfortunately, it is sometimes negated by the use of substitute additives, such as manganese, that may be just as hazardous as lead and that also emit particulates into the air.

MANGANESE

Manganese is a soft, silvery or reddish-gray metal, used not only as a gasoline additive but also in making dry cell batteries, paints, varnishes, dyes, inks, disinfectants, bleaches, fertilizers, matches, and in coloring glass and pottery. Early signs of manganese poisoning include headaches, insomnia, dermatitis, itching, decreased libido, and various respiratory problems. As the disorder progresses, it mimics the signs of Parkinson's disease: jerky, spastic muscle movements, shuffling walk, expressionless face, slow, slurred speech, and tremors.

A study conducted in 1978 by Morris M. Joselow and his associates[7] of ninety-eight inner-city children living near busy streets in Newark, New Jersey, highlights some of the potential hazards of air pollution caused by the substitution of manganese for lead in unleaded gasoline. From analyzing blood and soil specimens for lead and manganese, Joselow and his colleagues determined that levels of both substances in the test subjects and in the soil of the area showed a direct relationship to heavy traffic patterns: The closer a child lived to a heavily traveled street, the higher his or her blood levels of lead and manganese were likely to be. The average blood level for all subjects in the study for lead was 34 μg/dl and for manganese was 2.5 μg/dl, both significantly higher than normal.

As Joselow points out, we know far too little about the synergistic effects of lead and manganese in combination in the air to be allowing them to be emitted without further study and greater concern for their health effects. What little we do know about the synergistic properties of manganese alone is hardly reassuring: for one thing, it is a powerful catalyst in the formation of sulfuric acid rain.

BENZENE

Another pollutant commonly used as a substitute for lead in unleaded gasoline is one of the relatively few substances about whose ability to cause cancer in human beings there is no question. Benzene (C_6H_6), a member of the large chemical family of aromatic hydrocarbons, is a colorless, highly flammable liquid with a distinctive smell that easily vaporizes into a gas. Public exposure to it results from its wide use in paint removers, solvents, rubber cements, and numerous other industrial and household products, as well as gasoline.

Benzene has been known for some time to be a hazard to the central nervous system and a cause of aplastic anemia, a nonmalignant disease that causes a deficiency of white and red blood cells and platelets, but only in recent years has its use in solvents been unequivocally linked to the development of leukemia, especially among workers in rubber production plants, printing facilities, and shoe factories, who may be exposed to concentrations in the hundreds to thousands of ppm.

In the general public, benzene exposures rarely exceed the parts-per-billion level, and are ordinarily not risky. However, some health authorities note that the benzene fumes that collect at the top of a car's gas tank are often released in a whoosh when the cap is removed for a fill-up, and that ambient benzene levels as high as 250 ppb have been recorded at gas stations. This causes them to wonder if the recent trend toward self-service stations may be increasing the public's risk.

In occupational environments the current standard for benzene is 10 ppm, a level that some experts feel is ten times as high as it should be to give workers optimum protection. Unfortunately, it isn't likely to be lowered. The argument for a standard of 1 ppm has already been made by OSHA to the Supreme Court, which ruled that OSHA had failed to prove its case.

SMOG

There are two kinds of smog. The classic type, formed by the dust, soot, ash, and gases released by burning coal and other fossil fuels without adequate emission control devices, has been with us for hundreds of years, although the name "smog" (a contraction of *sm*oke-f*og*) wasn't coined for it until this century. We explore this type later in this chapter. The newer type, caused by the reaction of automobile and industrial emissions with the ultraviolet light in sunshine, is called "photochemical smog," or "photochemical oxidants," and has only become a serious problem in industrialized countries since the 1940s.

In a typical smoggy city like Los Angeles, New York, or Phoenix, Arizona, photochemical smog begins to form early in the morning, as early as five or six o'clock, as cars and industries begin to emit the first of the new day's load of nitrogen oxides, hydrocarbons, and particulates. The first pollutant that appears is nitric oxide (NO). As the day wears on, the steady addition of oxygen molecules to NO, a process hastened by the presence of sunlight, transforms it to NO_2. Nitrogen dioxide reaches its highest concentrations in the mid- to late-morning hours. Then a second round of chemical interaction begins, combining NO_2 with hydrocarbons (HC) and other volatile organic compounds, including benzene, and producing ozone (O_3), which reaches its maximum concentrations around the middle of the day. Usually smog levels will decline slightly in the afternoon, then peak again after evening rush hours, and drop to their lowest concentrations at night. Unless wind, rain, or other weather conditions dispel it, smog levels will continue to build up to higher and higher levels with each passing day. The worst smog conditions form in areas like the Los Angeles basin, ringed by mountains that block wind currents, on calm, sunny days when there is no rain or wind, during temperature inversions, which act like a lid on the air near the earth, keeping it from circulating upwards to distribute pollutants throughout a larger volume of air to lower their ground-level concentrations.

The main constituent of smog is ozone. This bluish, pungent gas, a close chemical cousin of oxygen, is also created naturally in the atmosphere by the electrical discharges we call lightning, and at very high altitudes by the simple interaction of sunlight and oxygen. Ozone is formed at low concentrations in many industrial processes involving high-voltage electrical arcs, such as mercury vapor lamps, X-ray machines, and some types of welding. Ozone has a number of useful applications. It has been employed as a bleach, drying agent, and a deodorizer (that fresh, clean smell in the air just after a thunderstorm is caused by traces of ozone). Ozone has also been used with varying degrees of success as a water disinfectant in swimming pools and water treatment plants.

It is ozone, more than any other compound, that gives smog its stinging, irritating qualities. The current ozone standard set by the EPA is 0.12 ppm, a level that is often exceeded in many cities. Levels up to and above 0.5 ppm are not unheard of. Fortunately, although the actual mechanical function of the lungs in healthy adults may be measurably affected by ozone at levels as low as 0.37 ppm, experimental data indicate that most adults in normal good health will not suffer conscious discomfort from ozone exposures below 0.5 ppm.[8] Sensitive individuals, especially those with asthma, emphysema, or other chronic lung conditions, *do* tend to react to lower levels, as low as 0.15 ppm, with coughing, wheezing, asthma attacks, chest pain, and a generalized feeling of illness. At very high concentrations, above 5 ppm, ozone can cause inflammation and congestion of the respiratory tract, choking and coughing with bloody sputum, and fatal pulmonary edema.

More recent research has revealed the strangely contrary nature of this intriguing gas. Although some evidence links ozone to hastening of the aging process, chromosome damage, and even to the formation of lung tumors, other studies suggest that ozone, at concentrations found in urban smog, may actually help *fight* cancer. As Frederick Sweet and his coworkers reported in August 1980,[9] ozone concentrations of 0.3 to 0.5 ppm slowed down the growth of human cancer cells from breast, lung, and uterine tumors by about 50 percent, leaving healthy tissues unaffected. At higher concentrations, 0.8 ppm or so, ozone almost completely stopped the growth of cancer cells, but it also began to retard the growth of normal cells. Sweet speculates that ozone may find a useful place as an adjunct in lung cancer treatment.

Ozone has yet another side to its many-faceted personality. Natural ozone in the stratosphere, some ten to fifty kilometers above our heads, screens us from excessive ultraviolet light from the sun. In 1974 researchers began to theorize that long-lived fluorocarbons such as Freon 11 and 12, widely used as propellants in spray cans since World War II, might be slowly ascending into the stratosphere, emitting chlorine gas as they decomposed, and steadily destroying the ozone layer. In December 1978 concern about this possibility led the EPA to ban most uses of fluorocarbon propellants. However, we are still manufacturing and using some 36 million kilograms of fluorocarbons a year as coolants in air conditioners and refrigerators and for a few other industrial purposes.

Investigators currently estimate that, along with nitrogen oxide emissions from fertilizers and jet exhausts, fluorocarbons may al-

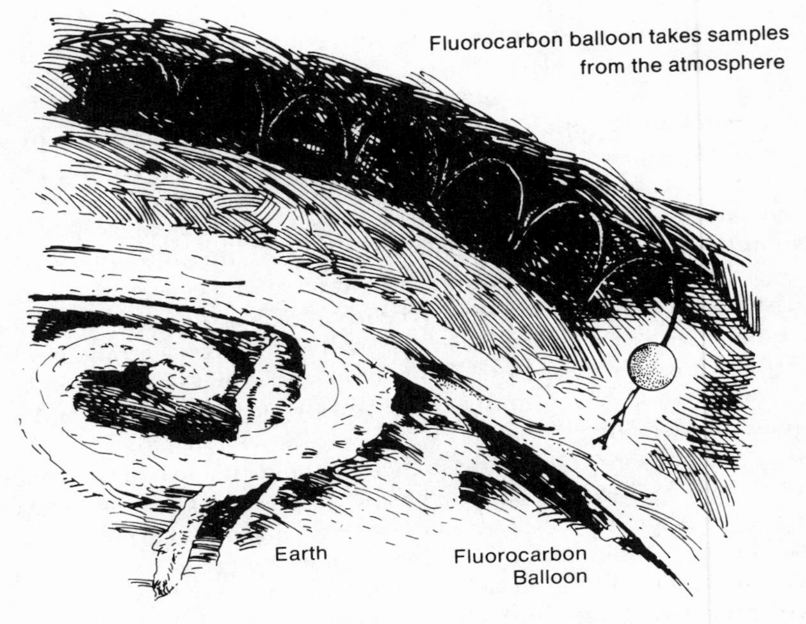

Figure 14

ready have depleted as much as 1 percent of the stratosphere's ozone, and predict that for every percentage point depletion in the ozone layer as much as 2 percent more ultraviolet light will reach the earth's surface, an increase that carries with it the potential to cause some two thousand to fifteen thousand extra cases of skin cancer a year. NASA is currently sponsoring a research effort involving high-altitude, instrument-crammed balloons to examine this theory more closely.

CARBON DIOXIDE

Although not ordinarily thought of as a particularly *toxic* air pollutant, carbon dioxide (CO_2) is arousing concern in some segments of the scientific community because it is thought that CO_2 in the atmosphere, caused mainly by the combustion of fossil fuels, may be leading to slowly increasing global temperatures (the "greenhouse effect"), which in turn may cause melting of polar ice caps, elevations of sea level that will flood many coastal areas, and destructive shifts in weather patterns that could turn some of the most productive farmlands in the world into deserts.

It is clear that global levels of CO_2 have risen slightly over the last century, from an 1860 average of 290 ppm to over 335 ppm today. (CO_2 does not begin to pose a direct threat to human health until its concentrations reach 30,000 ppm, at which point it can cause shortness of breath and headache. At 100,000 ppm CO_2 exposure causes visual disturbances, ringing in the ears, tremor, and loss of consciousness within one minute.) Each year the atmospheric concentration of CO_2 in the world as a whole increases by about 1 ppm, mainly due to the increasing use of fossil fuels in the industrialized nations. Currently, man contributes an annual load of somewhere between 8 and 14 billion metric tons of CO_2 into the air around the globe. Some of that is due to the burning of wood for heating and cooking, and incineration of wastes, but over 80 percent comes from the combustion of oil, natural gas, gasoline, coal, and other fossil fuels. By the year 2000, scientists predict that man will be pumping over 36 billion metric tons of CO_2 into the air each year, and the percentage derived from fossil fuels will have risen to 95 percent.

As with sulfur and nitrogen oxides, natural emissions of CO_2—measured in the trillions of metric tons—dwarf man-made levels. But nature's carbon cycles have been maintained in stable, dynamic equilibrium for millions of years. The additional CO_2 that nature generates from decaying plant and animal matter is offset almost perfectly by the CO_2-absorbing process of photosynthesis. The CO_2 that man produces, small though it is in comparison to that produced by nature, may unbalance the delicate harmony of this process.

However, though it is clear that CO_2 is slowly increasing, it is not clear that this is dangerous. In fact, global temperatures rose only until the late 1940s, and then began to drop, leading other scientists to postulate that we may be in more danger of a new Ice Age than a heat wave. Furthermore, there is some evidence that higher CO_2 levels might turn out to be beneficial rather than harmful. For example, we have known for some time that photosynthesis increases in the presence of higher CO_2 levels, a fact that commercial greenhouse operators have been quick to capitalize upon: To correct problems of slow growth and low yields that result from the depletion of natural CO_2 caused by many plants in an enclosed place, some owners have taken to pumping fossil-fuel-derived CO_2 into their greenhouses, raising its levels to as much as 3,000 ppm. The results have been impressive increases in yields and decreases in the length of time it takes for some crops to grow to maturity. We may also find that the new science of gene-splicing will allow us to engineer plant strains that thrive on an atmosphere rich in CO_2.

INDOOR AIR POLLUTION

There is no better illustration of the Sideslip Factor than the fact that the rush to save energy by insulating and weatherstripping older homes and building new structures as "tight" as possible has been accompanied by the appearance of a new environmental health hazard—indoor air pollution. Gradually, over the last five years, the realization has been slowly dawning in the scientific community and among the public that, since most of us spend most of our time inside (even kids spend more time indoors than out), the quality of interior air is as important as, if not more important than, the outside air quality, and the two are not identical.

Homes built before the energy crunch were generally quite "leaky"—air constantly infiltrated into them through cracks around doors and windows and joints in their construction. Ventilation and heating engineers usually assume that in these houses the air is naturally replaced about once every hour. In newer structures, deliberately built to conserve as much energy as possible by slowing down the infiltration rate, the air may not be replaced more than once every two to five hours, or even less, and it is in these buildings that indoor air pollution problems can develop.

Indoor air pollution can have an astonishing variety of sources, including gas ranges and space heaters, fireplaces, wood- or coal-burning stoves, cigarette smoking, household products like detergents, waxes, polishes, air fresheners, pesticides, glues, paints, hair sprays, oven cleaners, permanent press fabrics (clothes, sheets, etc.), synthetic fibers and protective coatings in curtains and carpeting, some types of furniture, certain kinds of insulation, cleaning activities like dusting and vacuuming, electrostatic air filters, and even the very materials a structure is made of, including particle board, plywood, brick, concrete, the ground it sits on, and the water that is piped into it. And, of course, whatever pollutes the outside air eventually becomes indoor pollution as well, especially in older, leaky buildings.

Any open flame—from gas ranges, gas space heaters, and fireplaces, for example—produces carbon monoxide and nitrogen oxides, both discussed in some detail above. So do cigarette smoking, gasoline-powered engines, and charcoal. Generally, CO accumulation from sources like gas ranges, heaters, and fireplaces is a minimal hazard, especially if they are properly vented and the vents are working correctly, but a car or power mower left running in a closed garage, or even a barbecue brazier used inside a poorly ventilated space, like a shed or camper, is an invitation to disaster, one that dozens of people ignore every year with tragic consequences.

NO_2 concentrations from gas appliances and heaters have a slightly stronger tendency to exceed safety levels. One study found that the highest indoor CO concentration reached was only 48 mg/m^3, only 20 percent higher than the EPA standard of 40 mg/m^3. But NO_2 concentrations were measured in the same study at 2500 $\mu g/m^3$, about a 500 percent excess over the 470 $\mu g/m^3$ maximum level recommended by the EPA.[10] Not enough is known about long-term exposure to low concentrations of NO_2 to say for sure what sort of health hazard, if any, these low levels may pose, but the way to prevent them is simply to make sure that ventilation systems, even if they're no more elaborate than opening a couple of windows to get a cross-draft, work and are used. Dwellings with all-electric appliances and heaters and no fireplace don't, of course, pose problems of NO_2 buildup at all. (Leaky, unburned natural gas, which is mostly methane plus a little ethane, propane, nitrogen, and butane, does not cause CO or NO_2 buildup either, but it can explode if it comes into contact with a spark or open flame, and it can displace enough oxygen in the air to cause simple asphyxiation.)

Dusting and vacuuming can double the level of respirable particulates in a structure, but the worst source of indoor particulate pollution is smoking. A study conducted in 1975 by Neville Lefcoe and Ion Inculet[11] concluded that smoking has a "tremendous" effect on levels of indoor particulates, increasing certain types as much as a hundredfold for as long as three hours after *one* cigarette has been smoked inside a residence.

One year after Lefcoe and Inculet's report Dr. Ralph Binder reported results of a study that strongly confirmed the particulate hazard caused by indoor smoking. Using portable air samplers, Binder studied the exposures of twenty children aged twelve to seventeen years to SO_2, NO_2, and respirable particulates over a typical twenty-four-hour period. The results showed that SO_2 and NO_2 exposures were low to moderate, but that "personal exposures to particulates were considerably higher than outdoor concentrations; in all but one subject they exceeded the present EPAAQS [EPA Air

Quality Standards]. . . . We were able to demonstrate a significant relationship between particulate concentrations and one of the several domestic factors that we measured: the effect of the presence of at least one smoker in the subject's house was striking."[12] As we explained above, probably the chief health hazard of particulates is that they make other pollutants more respirable.

Of greater concern to most health experts is formaldehyde (HCHO), a pungent gas widely used in color photography and papermaking, and in dyes, inks, and textiles. A liquid version, formalin, is a commonly used disinfectant and embalming solution. Inside residences, formaldehyde can be emitted by many sources—cosmetics, carpeting, permanent press clothes, drugs, and fiberglass insulation, among others—but high levels are invariably due either to particle board used in making furniture, cupboards, and partitions, or to a specific type of foam insulation in which formaldehyde is mixed with urea and injected into wall spaces to fill up empty air pockets. If manufacturing and mixing procedures are not followed meticulously in these cases, these products can give off excessive levels of formaldehyde and similar chemicals for months after installation, causing irritation complaints.

Low levels of formaldehyde (0.2 ppm) can cause a dry throat, irritation of the eyes, nose, and throat, and sinus complications. Chronic exposure to low levels of formaldehyde can cause insomnia, anxiety, and can trigger asthma attacks, or make them worse. Children, especially those with preexisting illnesses, are particularly sensitive to formaldehyde exposure. Short-term, acute exposures to high concentrations of formaldehyde can cause pulmonary edema and respiratory failure; fortunately, at high levels the fumes are so noxious few people can stand to be exposed to them long enough to receive a hazardous dose. There is also evidence from the National Academy of Sciences that low levels of formaldehyde can cause nasal tumors in rats and mice, but the implications of these findings for humans are still controversial.

In 1980 the Consumer Products Safety Commission urged that the use of formaldehyde in particle board, plywood, and insulation be banned. As of this writing, Massachusetts has adopted the commission's recommendation, and several states are considering doing so.

Perhaps the most serious indoor hazard of all, however, is a natural one, the radioactive gas known as radon. A derivative of radium, which is found very commonly in the soil, radon emits a kind of radioactivity that can cause lung cancer if inhaled at sufficient concentrations. The presence of respirable particulates in the air to which radon and its byproducts can adhere increases its hazards. Radon is also given off by masonry building materials—bricks, granite, gypsum, concrete—and by groundwater in some areas. (Radioactivity in general is discussed further in Chapter 8.)

Careful experimental studies have shown that radon concentrations will stay below hazardous levels in most areas in homes with at least one complete air exchange every two hours. (In a few locations with very high levels of radon in the soil, such as Grand Junction, Colorado, or Saskatchewan, Canada, radon may become a problem even in homes with high rates of air infiltration.) In extremely tight houses, with exchange rates as low as once every five hours, radon is a greater hazard, accumulating three to five times its levels in leaky homes.[13]

PASSIVE SMOKING

That smoking is hazardous to smokers should be no surprise to anyone. And as we have just seen in the preceding section, there has been good evidence since 1975 that smoking is a primary cause of at least one type of indoor air pollution, respirable particulates. But does the smoke from either the burning cigarette (or pipe, or cigar) itself—known as "sidestream" smoke—or the smoke that is exhaled back out into the air from a smoker's lungs—known as "second-hand" smoke—actually hold the potential to cause serious diseases in those who must breathe it simply because they live or work in the same place as smokers?

New attempts to answer this question surfaced in 1981, with inconclusive results. For many decades Japanese health researchers, noting that lung cancer was rising at a rapid pace in Japanese women even though only 15 percent of them smoke, have speculated that it may be because so many Japanese men (about 75 percent) do. In 1981 the results of a fourteen-year study of nearly one hundred thousand married nonsmoking Japanese women strongly suggested that one reason a higher percentage than expected died of lung cancer was that they were married to men who smoked heavily, more than a pack a day. They were, in other words, victims of passive smoking.

According to Dr. Takeshi Hirayama, Chief of the Epidemiology Division of the Japanese National Cancer Center Research Institute in Tokyo, "continued exposure to their husbands' smoking increased mortality in nonsmokers up to twofold."[14] There was also some evidence that passive smoking may have been linked to fatalities from emphysema and asthma in these women. Dr. Hirayama concluded that the health risks of passive smoking were a very high one-third to one-half those of direct smoking.

Six months later, however, Hirayama's findings were hotly disputed by the Tobacco Institute and the American Cancer Society (ACS). The Tobacco Institute claimed to have found methodological errors in the Japanese study that undermined its validity, and the ACS's Lawrence Garfinkel pointed out that not only had a study he conducted of some 175,000 female nonsmokers failed to turn up any evidence of greater risk from marriage to heavy smokers, but that, in the United States at least, marriage was only one of many possible exposures to sidestream and secondhand smoke. Neither study, Garfinkel argued, had really answered the question about passive smoking.

Somewhat overlooked in this controversy about spouses was the fact that numerous studies over the past few years have, with much less ambiguous evidence, pointed to the fact that it is the children of smokers who really suffer from their exposure to sidestream and secondhand smoke. From their first breath (and even earlier for fetuses whose mothers are heavy smokers) into their teen-age years, children whose parents smoke have increased rates of coughing, bronchitis, asthma, emphysema, and allergic-type reactions, as well as detectable decreases in lung function. It is clear that passive smoking is, at the very least, an irritant for any nonsmoker, and may be much more hazardous than that. But for children and adults who already suffer from asthma, emphysema, chronic bronchitis, and other respiratory diseases, this may be bad enough.

THE RETURN TO COAL

As the costs of oil-generated electrical power continue to rise, and nuclear power becomes ever more entangled in construction cost overruns and safety issues, coal is slowly becoming a competitive fuel source once again. Innovative research on novel, highly experimental energy sources like fusion and the liberation of hydrogen from water will no doubt continue but it will be decades, if ever, before we see the practical fruits of these projects.

Coal is nothing more than ancient solar power, in the form of decayed plant matter, densely compacted to various degrees of hardness. Chemically, coal is a mixture of carbon, nitrogen, sulfur, hydrogen, oxygen, and water formed naturally as plants die and are slowly covered with layer after layer of still more decaying organic matter, soil, and rock. Over millions of years bacterial action and the steady pressure of these layers welds what was once fragile vegetation into peat, then into a low-grade coal known as lignite, and finally into rock-hard sub-bituminous, bituminous, and anthracite coals, the kinds most commonly used today.

Coal was used as a fuel in China before the birth of Christ, and found some applications in medieval metallurgy in Europe. But it wasn't until wood and charcoal began to become scarce in the sixteenth century that it first came into widespread use as a fuel in England, where, some four hundred years and numerous public health tragedies caused by coal pollution later, the debate over how best to manage its noxious fumes and smoke still goes on today.

Coal provoked dismay in England early in its long career there. As an irate John Evelyn wrote in 1661:

> This is that pernicious Smoake which sullyes all her Glory, superinducing a sooty Crust or furr upon all that it lights, spoyling the moveables, tarnishing the Plate Gildings and Furniture, and corroding the very iron-bars and hardest stones with those piercing and acrimonious Spirits which accompany its Sulphure . . . It is this horrid Smoake which obscures our Churches, and makes our Palaces look old, which fould our Clothes, and corrupts the Waters, so as the very Rain, and refreshing Dews which fall in the several Seasons, precipitate this impure vapour, which, with its black and tenacious quality, spots and contaminates whatsoever is expos'd to it.[15]

Despite Evelyn's outrage, shared by most of his contemporaries, by the late seventeenth century England was the world's foremost producer of coal. A century later the invention of the steam engine, with its intrinsic demand for a more compact, higher-energy fuel than wood, triggered the Industrial Revolution and sent an already strong demand for coal skyrocketing.

In North America the same pattern emerged, a century deferred. The eighteenth century was an age of wood and charcoal here, preferred over coal because of their cleaner burning properties. By the nineteenth century, however, coal was in the ascendancy in the United States, and by the century's end America had edged out Great Britain as the world's leading producer of coal.

As urbanization took hold here and in Europe, greater concentrations of people, reduced air circulation, and higher average temperatures in cities all combined to cut down on natural processes of pollution dispersal in the air, and large-scale pollution accidents began to occur in coal-burning areas with monotonous regularity. In 1909 over a thousand people died in Glasgow in a series of coal smog incidents. In 1948 twenty people died and hundreds more were injured in Donora, Pennsylvania, in a coal smog tragedy. Four years later, in December 1952, the greatest coal smog disaster of all, one that finally mobilized successful efforts to pass tough pollution control laws, occurred in London itself.

December is a cold month in England, and in many homes that year coal-fueled heaters were kept burning all night long. Early Friday morning, December 5, a thick, choking coal smog descended on the city and stayed there for the next four days. Curiously enough, SO_2 concentrations never rose above 1.34 ppm, a level found commonly today in heavily industrialized areas of the world. But combined with the presence of high levels of respirable particulates in the air, these SO_2 concentrations proved lethal. By the end of the following week, on December 14, it was clear that the smog had been a record killer. The death rate for London County, normally well below 1,000 per week, jumped to 2,484 and remained well above 1,000 per week until the end of the month.[16] Most of the excess deaths were caused by exacerbation of previous respiratory problems in the very young and old, and it is now thought that the adsorption of SO_2 onto coal particulates (called "fly ash") was the major hazard. As a result tuberculosis deaths quadrupled, pneumonia mortalities rose fivefold, and flu fatalities climbed to seven times their normal level. In all, some 4,000 extra deaths were ultimately attributed to this one event.[17] Four years later Parliament passed the first in a series of tough air pollution laws prohibiting, among other things, the burning of coal in open grates, and requiring industry to begin adopting more effective types of coal-emission control devices.

There have been radical advances in the technology of coal pollution control in the last thirty years, and we now have much tougher laws in this country to limit what can be emitted from coal-fired plants, but incidents like the one described above, as well as the fact that our own air pollution laws need strengthening, raise critical questions about our return to coal.

The United States possesses some 30 to 35 percent of the world's proven coal reserves, far more than any other country—the Soviet Union has 25 percent, China has 15 percent, and the rest is scattered around the globe. If we continue to consume oil at present rates, the world's supplies will run out by the middle of the next century, but there is enough energy trapped in the world's coal to meet our needs for the next three hundred years at current rates of consumption.

Most of our coal is located in either the Appalachians or the Rocky Mountain Basin, a vast region stretching along the spine of the Rockies and into the Great Plains, one that includes large beds of coal in Wyoming, Montana, the Dakotas, Utah, Colorado, New Mexico, Texas, Oklahoma, Kansas, Iowa, Illinois, and Indiana. In the future the major growth area in coal mining will take place in these states, which produced only 25 percent of our coal in 1980 but will be producing nearly half of it by the end of the decade.

The area where the return to coal is at its strongest right now is the generation of electrical power. Utilities have slashed their use of oil significantly in the last few years—by 26 percent in 1980—and are turning more and more enthusiastically to coal. Over half of our electricity is now generated in coal-fired plants; almost 75 percent of the coal burned in the United States is used for this purpose. Over the next decade some 275 coal-fired electrical generating plants are scheduled to be built around the country. Utilities are also actively exploring ways to convert coal into "synfuels"—gases that can substitute for natural gas and liquid fuels that can replace gasoline.

Despite all this our return to coal has been surprisingly slow. We use only slightly more coal today than we did twenty-five years ago. Demand has stayed at low levels, increasing only a few percentage points per year. Our 1979 consumption of coal, 671.5 million metric tons, equals only about one-fifth of our energy needs. We have the capacity to produce another 90 million metric tons a year right now, if the demand for it existed, and that 90 million metric tons would replace some 400 million barrels of oil for which we are currently paying well over $10 billion.

Part of the delay in the return to coal is caused by the inescapable fact that coal is a dirty fuel. What was true in seventeenth-century England is still true today, despite phenomenal advances in the science of controlling air pollution. Coal combustion releases all the same pollutants that any other carbonaceous fossil fuel does—carbon monoxide, carbon dioxide, sulfur and nitrogen oxides, and particulates—but the two substances that are of greatest concern are sulfur dioxide and fly ash. Fly ash was the primary cause of the black, sooty grime that covered cities in Victorian England and was known in many cities in the eastern and midwestern United States, especially St. Louis and Pittsburgh, as recently as the 1930s. It was also, as we have just seen, a significant factor in the 1952 London smog catastrophe.

Systems for controlling coal pollution include everything from using coal with low-sulfur content, washing the coal before it is burned, and keeping the air supply to the combustion chamber as low as possible, to the use of substances that absorb sulfur (dolomite, lime, limestone) and prevent it from reacting with oxygen to form SO_2, as well as more complex and expensive alternatives like activated charcoal filters, electrostatic precipitators, scrubbers, and systems that capture and incinerate pollutants as they are released from exhaust stacks.

Today, fly ash control is the coal industry's success story. In most large coal-fired plants, electrostatic precipitators are used to control these particulates. Electrostatic precipitators are simply devices that establish a high voltage electrical field, so that when a stream of gases containing particulates enters the field, the particulates receive a strong electrical charge and as a consequence are attracted to a second electrode with the opposite charge, known as the collector electrode. The particulates then collect near the second electrode, clump together, become steadily heavier, and eventually settle or drain downward into a hopper or sump below. Electrostatic precipitators work very well. Some clean out up to 99.5 percent of the fly ash passing through them, and most are able to keep the gas stream cleaner than called for in air quality standards and local ordinances.

Gases of any kind, especially oxides of sulfur, are usually removed by devices called scrubbers, a generic name for any system that separates a gas from a stream of exhaust by bringing it into contact with a liquid into which the gas is driven by chemical reaction or by simple diffusion. (Diffusion is the term for the fact that a gas tends to move *from* an area of relatively high concentration *to* an area of low concentration, e.g., from a heavily contaminated exhaust stream *to* an uncontaminated liquid.) There are a number of ingenious ways to bring the gas and liquid into close contact in scrubbers: by bubbling the gas through a tank of the liquid, by spraying the liquid into the gas stream, or, perhaps most common, by forcing the gas into the bottom of a column filled with perforated plates or other packing material and the liquid into its top. The gas, which is hot, rises, and the liquid falls. When they meet, they combine, the pollutant reacts with the liquid, and is exchanged into it. The plates or packing material in the column serve the purpose of adding the abundant additional surface area needed for this exchange process to take place.

Scrubbers are probably the most efficient way of cleansing gases from an exhaust stream, but they are not as efficient as electrostatic precipitators. Scrubbers generally cannot remove more than 95 percent of the SO_2 or SO_3 from a coal plant's emissions, so sulfur remains one of the more problematic of coal emissions. Furthermore, scrubbing systems, like electrostatic precipitators and any other complex pollution control equipment, are expensive and their costs add to the price of the delivered electricity generated by coal combustion.

Although coal-fired plants emit only one-third the volume of nitrogen oxides that they do of sulfur oxides, these may be the most intractable pollutants of all, simply because, as of this writing, no effective, practical system for removing them had yet been developed for the U.S. market.

REGULATING AIR POLLUTION

The first national legislation aimed at combating the growing problems of air pollution appeared in 1955 in the form of the Air Pollution Control Act, which mandated research into the health effects of urban smog, primarily from auto emissions. Eight years later the legislative basis for all of our subsequent pollution control efforts, the Clean Air Act of 1963, was passed. Endless efforts to refine it led to repeated amendments in 1965, 1967, 1970, 1977, and 1981.

The act places the burden of air pollution control on the states and the regional agencies that oversee the nation's 247 air quality control regions. Throughout the 1970s the EPA played a research, advisory, and occasionally a watchdog role, reviewing state implementation plans, checking to see that local air quality boards were monitoring polluters effectively, and only as a last resort moving to impose penalties on areas and states that did not implement control programs quickly enough.

Early in the decade the agency developed primary standards for these seven major air pollutants:

- CO 10 mg/m^3 (9 ppm) average over eight hours (no more than 40 mg/m^3—35 ppm—average for any one-hour period)
- SO_2 80 $\mu g/m^3$ (0.03 ppm) average per year (no more than 365 $\mu g/m^3$ average for any twenty-four-hour period)

NO_2	100 $\mu g/m^3$ (0.05 ppm) average per year
Particulates	75 $\mu g/m^3$ average per year (no more than 260 $\mu g/m^3$ average for any twenty-four-hour period)
O_3	240 $\mu g/m^3$ (0.12 ppm) average per hour
HC	160 $\mu g/m^3$ (0.24 ppm) average per three-hour (six to nine A.M.) period
Lead	1.5 $\mu g/m^3$ (0.006 ppm) average per three-month period

As well as primary standards, intended to safeguard health, the EPA also established secondary standards to protect property and the environment—crops, wildlife, soil, water, and so forth. Separate standards were set for especially hazardous air pollutants—asbestos, mercury, vinyl chloride, and beryllium—and new ones are being considered for benzene, radioactive compounds, and inorganic arsenic.

The overall impact of the act has been to inventory polluters, monitor their emissions, require them to develop ways to control their pollution, inspect them to make certain that they are implementing these systems, and set deadlines for attainment of the primary and secondary standards. The act also set up a classification system designed to prevent future deterioration of air quality in areas that were already pollution-free, especially parks and wilderness regions, usually by prohibiting future industrial development in or near them.

The process of cleaning up the air has been a complex one, but some progress has been made. Many pollutant levels have dropped significantly in the last decade, due directly to the pollution control measures that the Clean Air Act requires. The biggest problems have been with carbon monoxide and nitrogen dioxide from motor vehicles, and ozone concentrations in especially smoggy urban areas. Only on rare occasions has the EPA been forced to exercise its punitive powers, which usually involve blocking of federal funding for highway development and water-treatment plant construction.

Progress has been painfully slow, however. Less than half of the air quality regions in the country are fully in compliance with the act. Deadlines have been repeatedly postponed, first from 1975 to 1982, and now to 1987, mostly because the automobile manufacturers have consistently dragged their feet, putting up tremendous resistance to any kind of emission control program at all from the very start, despite ample evidence that controls need not add exorbitant costs to new car prices, do not degrade a car's performance, and do not reduce its fuel efficiency.

At the same time, some trends are appearing that have the potential to make the problem of motor vehicle pollution much worse. Because they get better mileage than gasoline-powered cars, and because diesel fuel has retailed at slightly lower prices than gasoline in the past, diesel cars and trucks have been growing in popularity in the last ten years. The U.S. Department of Transportation predicts that by 1985 one out of every four new cars or trucks sold in the United States will be diesel-powered, and the National Academy of Sciences thinks that by the year 2000 there will be fifty times as many diesels on the road as there are today.

But diesel engines are even worse air polluters than ordinary gasoline-fueled cars equipped with catalytic converters. Emissions of carbon monoxide and hydrocarbons are similar, but diesels pump out more nitrogen dioxide, ten times as much sulfur dioxide, and, worst of all, thirty to a hundred times as much particulate matter as do their catalytic-converter-equipped counterparts. Many of these particulates are of respirable size, and some, like benzoapyrene, are known carcinogens. Many others (there are thousands of different types of particles in diesel exhaust) have not yet been fully analyzed or even identified yet. We do know from studies of diesel exhaust concentrations and cancer incidence in Switzerland, where every third car is a diesel, that there is a significant connection between heavily traveled traffic routes and elevated cancer rates.

The Reagan administration, vowing to encourage industrial development, is at this writing proposing a set of amendments to the Clean Air Act that could destroy the progress that has been made. These include allowing twice as much pollution from motor vehicles as is now permitted, dropping the deadlines for compliance with air quality standards, eliminating requirements that substandard areas show progress toward improving their air quality, weakening the EPA's enforcement role, allowing new polluters to use less expensive and less effective control techniques than the best available, eliminating restrictions against new polluters in relatively clean areas, and dropping current requirements that smog devices be regularly checked and maintained.

Industry, especially the car manufacturers, is delighted with these proposals, but environmentalists, health professionals, and the public as a whole—if polls mean anything—are not. These setbacks come at a time when new medical evidence is showing that many pollutants, sulfur dioxide in particular, have the ability to cause serious breathing difficulty for the 8 to 15 million asthmatics in this country at much lower levels than previously thought, as low as 0.5 ppm, well below levels commonly found in many polluted cities today. One of Reagan's proposals would make it much easier for coal-fired plants to emit more carbon monoxide, nitrogen oxides, and sulfur dioxide into the ambient air. Many plants that would currently be required to use scrubbers to remove these gases would not be under Reagan's revisions. If the lesson of the 1952 London disaster means anything, this may be a prescription for disaster, not economic recovery.

5

HAZARDS IN THE WATER

WATER: AN EVERYDAY MIRACLE

Dozens of times a day those of us who live in the industralized nations of the world enjoy a blessing denied to 75 percent of the world's population: abundant supplies of clean, cheap water. We sip it, splash in it, spray it on our lawns, and rarely give it a second thought. But water is even more essential to life on earth than air; no known organisms can live without it.

Chemically, water is H_2O—two light hydrogen atoms joined to one much heavier oxygen atom. It is the most abundant substance in the 50 trillion cells that make up the human body, accounting for 70 to 85 percent of the protoplasm that every call is made of. Water is more than just essential to good health—it is fundamental to life. It is a critical reagent in the constant biochemical reactions that take place in our cells, as well as the perfect solvent and transport medium for all the chemicals—sodium, potassium, magnesium, phosphate, and others—that cellular life depends on. Every second relatively huge quantities of water are passing in and out of every single one of the body's cells, a process so well regulated that the cell's net volume of water never changes.

We are creatures of water. At birth 75 percent of the human body is water. By age ten that percentage drops to 57 percent, where it remains for the rest of our lives, unless we get fat. Then, contrary to popular myth, we have even less water in our bodies, as little as 47 percent.

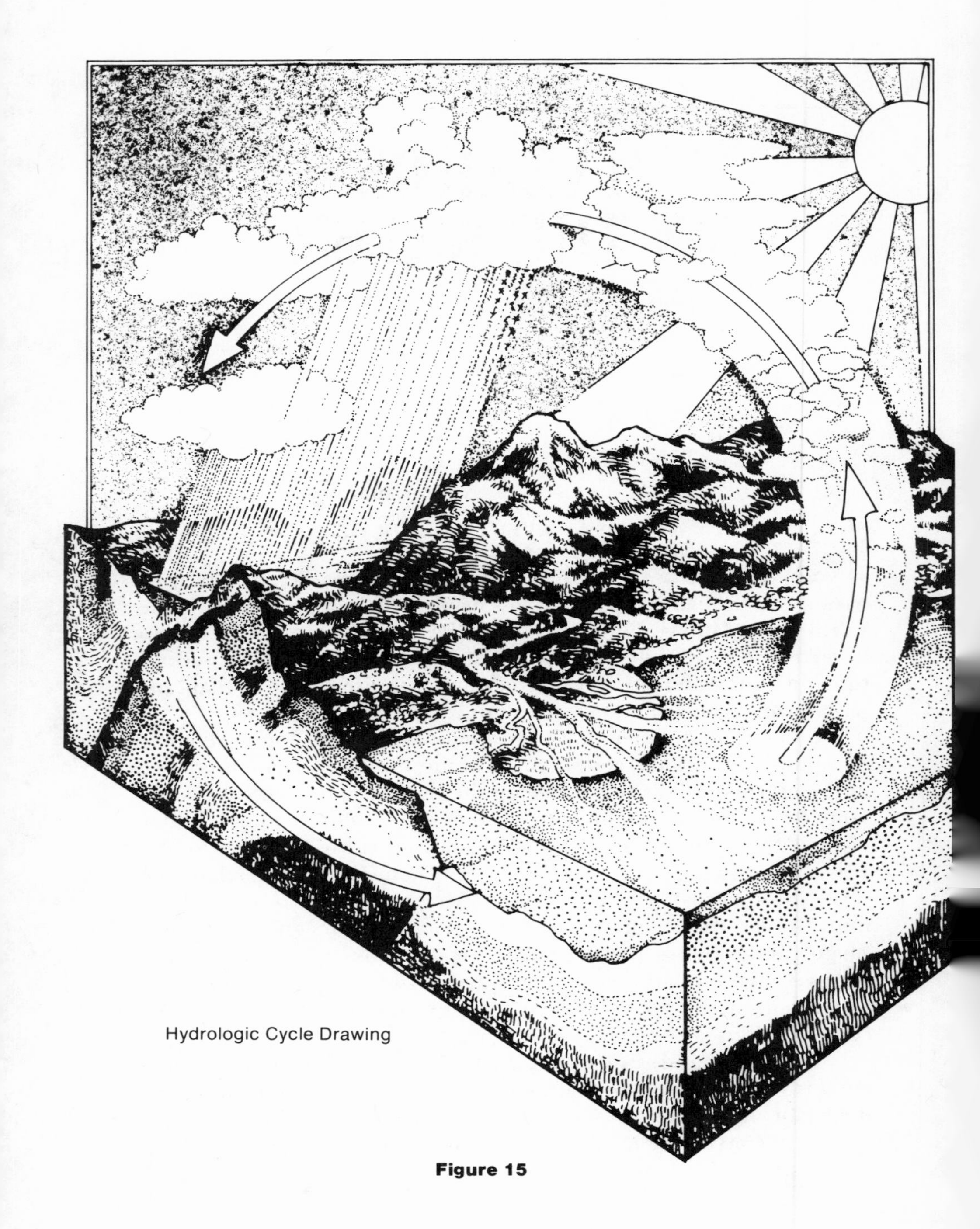

Hydrologic Cycle Drawing

Figure 15

The planet we live on is as watery as we are: 70 percent of its surface is covered with water. Between the earth's surface and the thin membrane of air that surrounds it water constantly circulates in a vast solar-driven process known as the hydrologic cycle: the sun's heat evaporates trillions of liters of water each day and draws them up from the surface of the land and sea into the atmosphere, where it slowly cools. Eventually, condensing and combining with other molecules of water, it gains enough weight to fall back to the earth's surface as precipitation. Each day the hydrologic cycle brings about 15 trillion liters of water in the form of rain, snow, hail, and sleet to the surface of the United States alone.

WATER UNDER THE EARTH'S SURFACE

When we think of water supplies, we are used to thinking of surface sources—rivers, streams, lakes, and man-made reservoirs. About half of us do get our water from surface sources, but the rest of us draw it from underground. Hydrologists calculate that within eight hundred meters of the surface in the United States there is about twenty times as much water as there is in all of our surface sources combined. Of the 900 billion liters of water we use every day in the United States, nearly half comes from underground sources; about 70 percent of this amount is used for agricultural irrigation.

Water penetrates under the earth's surface because the earth is more like a sponge than a skin. One-third of the water that falls on it soaks into it rather than evaporating or running off into nearby rivers or streams. Water that sinks underground eventually enters geological formations of sand, gravel, or rock that hold it in place, and there it stays until it is tapped into and drawn back up to the surface. These formations are called aquifers.

Since we can't see them and they don't have definite boundaries, aquifers are hard to measure, but we know that most of the country lies over some kind of groundwater source, and that certain aquifers are huge. The 400,000-square-kilometer Ogallala aquifer, for example, is as large as the states of Oklahoma, Kansas, Nebraska, and part of Texas that it lies beneath.

Aquifers are constantly being replenished by seepage from the surface, but this process is very slow. The water in some aquifers is thousands of years old. In areas where demand for groundwater outstrips the recharging rate, aquifers are being depleted, just like oil wells. One of the major problems we face today is how to replenish dwindling groundwater sources without exposing them to pollution from the surface.

A DECEPTIVE PLENTY

Despite the awesome volume of water that circulates every day through the hydrologic cycle, and the even greater amounts that are captured more or less permanently in the earth's underground aquifers, mankind faces the bleak prospect of increasingly severe shortages of water in the near future. By the year 2000, Americans will be using about 4 trillion liters of water a day, 50 percent more than will be available from natural sources.

In part the reason is the sheer growth of the world's population. The amount of water falling to earth each day remains more or less constant, but the number of people using it—and polluting it—is not. By the end of the century the earth's population will have grown by more than 50 percent from today's levels. Most of that growth will take place in the developing nations of the world, which already have the most trouble providing their citizens with clean, safe water.

Furthermore, the highly industrialized condition that most of these nations aspire to is an exceptionally thirsty one. For simple physical survival, a human being must have about two liters of water a day. Cooking usually adds another five to six liters. Household activities like showering, dishwashing, laundry, toilet flushing, and watering the yard may add anywhere from three hundred to six hundred liters or more. But the indirect water costs that members of a highly industrialized country incur add thousands of liters to that total. For example, making the steel that goes into the typical car requires enough water to add fifty to sixty liters a day to its owner's water consumption figures if the car is kept for ten years.

Moreover, in their desperate rush to industrialize as rapidly as possible, many developing nations are destroying the same natural watersheds they will need to meet the growing demand for water in their countries. In Brazil and parts of Southeast Asia, for example, huge chunks of rainforest are being burned or logged, to clear the land for more marketable crops, to sell off the increasingly valuable timber growing on it, or both. But all too often the delicate topsoil that has slowly accumulated in these ancient forests is left with no protection against rains and erosion, and the result is flooding, heavy deposits of silt in existing waterways, and eventually the transformation of a verdant jungle, drenched with usable water, into a near desert from which no moisture can be gleaned at all.

How will our need for more water be met? In part by conservation efforts: Each one of us will have to learn to use less so that more will be available for all. But most of it will have be met by recycling. Germany, Israel, and a number of other nations have already begun developing sophisticated and expensive water recovery plants, usually for crop irrigation rather than direct human consumption. In the United States, systems are just beginning to come onto the market that are capable of recycling residential water for individual households. Since recycling can, as we pointed out earlier, cause higher concentrations of pollutants already in the water, we will also have to learn to be even more careful about letting our water supplies become polluted in the first place.

SUN AND SOIL: NATURAL WATER PURIFIERS

Both the hydrologic cycle and the seepage of water through the earth's top layers of soil have remarkable powers of natural decontamination, up to a point. Because it converts water to vapor, the hydrologic cycle acts like an enormous distillery: In the same way that water is purified by being boiled into steam and then recondensed, so the bonds between water molecules and pollutants are broken when the sun turns surface water into vapor and draws it up into the atmosphere. When it recondenses, it is as clean as the distilled water you buy in a store—and if it didn't inevitably pass through contaminated air on its way back to earth, it might stay that way.

Similarly, water that seeps through the first few inches of the earth's crust is subjected to a number of physical and biological processes that cleanse it. The soil mantle acts like a vast membrane: some substances—water molecules—are allowed through; others are filtered out.

Mostly this happens because the top layer of soil is biologically very busy. Many of the impurities in polluted water are taken up by plants or consumed by microorganisms. Others adsorb to particles of inorganic matter in the soil. Waterborne bacteria and viruses that might cause disease tend to find that the soil is a hostile environment, one in which they can't survive for long.

Historically, the main benefit of this natural soil-filtering system was that groundwater was pure enough to drink right from the well or spring, without additional treatment. However, many of the synthetic substances that industrial societies are so proficient at dispersing throughout the environment are immune to these processes of natural biodegradation and decontamination, and pass through the earth's soil filter with ease. DBCP is one such substance. We describe others in this chapter.

Groundwater resources are particularly vulnerable to pollution of any kind because they are static. Whereas a river may have a flow rate of several meters per minute, which helps to dilute and quickly flush away pollutants, groundwater barely moves at all, as little as a few meters a *year* in many places. Thus toxic pollutants that enter an aquifer tend to remain there, concentrated in one area.

Furthermore, a polluted aquifer is nearly impossible to decontaminate. If the contaminants are toxic and persistent, the only practical solution to severe aquifer contamination is to stop using its water, and this of course places an additional demand on other water supplies, already overburdened in many parts of the country. Water from a polluted aquifer can be drawn up to the surface and then treated just like surface water, but the process is usually prohibitively expensive, whether the user is an individual homeowner or a municipal water company.

CONVENTIONAL WATER POLLUTANTS

Conventional water pollutants are substances that come primarily from nonindustrial sources and are highly biodegradable. They are a traditional problem in any human society, highly industrialized or not. They are also substances that we know a great deal about and for which we have developed successful control methods. Here are a few major examples: fecal coliform bacteria, nutrients, nitrates, and sediments.

COLIFORM BACTERIA AND THE BROAD STREET WELL

The fact that animal wastes can pollute drinking water has undoubtedly been known since the dawn of human civilization, but it was only within the last century or so that the reasons for this fact were explored scientifically. The first causal epidemiological connection between human feces and disease was made by Dr. John Snow, an English physician. In 1854 Dr. Snow unraveled the mystery of a

London cholera epidemic by observing that the common link between all the victims in the Broad Street area was that they obtained their drinking water from the same well. Even those who lived some distance away, Snow noted, had taken some pains to have water from the Broad Street well delivered to them because they preferred its taste. Further investigative work on Snow's part revealed that a sewer pipe passing close to the well had developed a leak and was infecting the water with human waste. Despite this early insight it took another thirty years before the bacteria in human fecal matter that can cause cholera and other diseases were isolated and identified by a German scientist, Theodor Escherich.*

Today, we know that coliform bacteria make up something like one-fifth to one-third of the average person's wastes, and that billions of them are released into sewage systems every day. We also know that some of these bacteria can be remarkably toxic. There are instances on record where the contamination caused by *one* individual's feces has caused thousands of cases of diseases like cholera, typhoid fever, and gastroenteritis. The single most effective weapon that we have yet devised against this threat is chlorination, but as we point out later, chlorination may pose health hazards of its own.

NUTRIENTS

Nutrients like phosphorus, nitrogen, and carbon are essential to aquatic life. However, too much of a good thing, in the form of an excess of these nutrients in waterways, can cause a paradoxical kind of pollution known as "eutrophication," from the Greek word for "well nourished." In bodies of water like ponds and lakes, which lack the flushing action of the high flow rate of rivers and streams, an excessive supply of nutrients can lead to the development of thick growths of algae and other aquatic flora along the water's surface. This prevents sunlight from penetrating into the water's depths and inhibits the growth of essential plants there. Accelerated algae growth is, of course, inevitably followed by accelerated algae death, and by the buildup of layers of dead organic matter on the bottom. As this matter decays, it gives off noxious odors, clouds the water, and uses up the precious dissolved oxygen that most aquatic life depends on.

Many of the nutrients that cause eutrophication originate as agricultural fertilizers that run off into surface waters from fertilized

* After whom E. coli (*Escherichia coli*) is named.

fields. They may also come from household wastes, especially laundry detergents containing phosphates, although use of this additive is now outlawed or severely restricted in many places. Unfortunately, most municipal sewage treatment plants are far more effective at removing other kinds of pollutants than nutrients, and in some cases they may even help convert nutrients into a persistent mineral salt form that is more easily absorbed by aquatic flora.

Lake Erie, long the sickest of the Great Lakes, has had to contend with severe eutrophication problems along its shores for some years, primarily caused by sewage effluent from Detroit. The problem is beginning to be resolved, but estimates of the future cost of halving the phosphate level in the effluent run to about $100 million a year more than is available for such treatment.

NITRATES

One nutrient, nitrogen, can also cause other kinds of trouble. Nitrates occur naturally in water at low levels. Most of this "background" level of nitrates is caused by ammonia, a form of nitrogen found in animal wastes and sewage. However, in areas where farmers have come to rely on heavy applications of artificial fertilizers—most of which contain nitrogen in the form of urea—as the easiest way to replenish nitrogen in the soil when it becomes depleted, nitrate buildup may become a problem.

High levels of nitrates in drinking water can be toxic to human infants and some kinds of newly born livestock. According to the U.S. Public Health Service, levels above 10 mg/liter in water can cause methemoglobinemia, otherwise known as "blue baby," a condition resulting from inadequate oxygen in the blood. Nitrate levels as high as 40 mg/liter have been found in some parts of the Midwest, usually due to the heavy use of nitrogen fertilizers on corn crops. Correcting the problem is simple: Farmers are advised to cut down in the amount of fertilizer they use and to use polluted water for irrigation only, since plants will absorb the nitrates, thus lowering their concentrations in the polluted water supplies.

SEDIMENTS

Sediments are the particles of sand, grit, and other inorganic matter that flow into waterways from mining sites, construction sites, logging sites, and urban runoff. Severe sedimentation can physically fill in small ponds and lakes that aren't protected with appropriate embankments and settling ponds, making expensive dredging operations necessary if they are to be kept open. Sediments can also damage the equipment in sewage treatment plants. Finally, like too many algae, excessive sediments suspended in water, a condition known as "turbidity," can block out the sunlight that aquatic organisms need to survive.

Mining activities are one major cause of sedimentation problems. Another source in cities and suburbs is the construction of new homes and roads, especially in hilly areas. Construction work tends to tear up the natural ground cover that keeps hillsides from washing away when it rains. At the same time it also tends to disturb the soil, making it more susceptible to erosion. As a result, large amounts of dirt, sand, and gravel can be washed away by natural runoff after heavy storms. Moreover, the slick road and roof surfaces that replace the natural ground cover aggravate this problem.

ACID RAIN

Acids can enter waterways from a number of sources, including industrial effluents and active or abandoned mine shafts that fill up with rainwater and then leach natural mineral acids into local waterways. In recent decades, however, the major concern about acidic waterways has focused on the problem of acid rain, caused in large part by the combustion of fossil fuels, which, as we explained in the previous chapter, release oxides of sulfur and nitrogen into the air, where they can mix with water vapor to form sulfuric and nitric acids.

Rain is naturally slightly acidic, and there are other substances besides sulfur and nitrogen oxides that increase its acidity. Carbon dioxide, for example, which enters the atmosphere from numerous sources, including animal expiration and the burning of wood and fossil fuels will lower the pH of rain to 5.6 by itself.* On a worldwide

*The pH of a body of water is a measure of its relative acidity. Neutral water has a pH of 7; higher numbers denote water that is more and more alkaline; lower numbers indicate water that is more acid. Each step of one number represents a ten-fold change in pH; water with a pH of 5 is ten times as acidic as water with a pH of 6, and a hundred times as acidic as water, with a pH of 7.

As water grows progressively more acidic, a number of problems can arise. Acidic water can corrode the pipes it is transported in, causing maintenance problems and basis 90 percent of the nitrates and 70 percent of the sulfates in the air come from natural rather than man-made sources. But, as we pointed out in the previous chapter, nature tends to be able to keep its own pollutants in balance, and it is the industrial emission of sulfur and nitrogen oxides, primarily from fossil fuel combustion, that now threatens to overwhelm natural processes of decontamination. We know that since the 1970s, sulfur dioxide emissions in the world have increased by 35 percent, and that rain has become more acidic.

The problem of acid rain first came to the attention of scientists in Northern Europe more than twenty years ago. Observing that lakes in southern Sweden and Norway were losing their fish populations, researchers began monitoring pH values and air samples and concluded that prevailing atmospheric currents were carrying acid mists caused by coal-fired plants in Great Britain across the North Sea and depositing them in Scandinavia. In 70 percent of some fifteen hundred lakes analyzed in Norway, pH values were below 4.3 and they contained no fish at all. Ironically, these lakes are crystal clear and quite beautiful.

Alerted by these findings, scientists in other countries around the world were soon reporting increased acidity in their waterways. In North America the problem is particularly acute in the Northeast, including large areas of both Canada and the United States. Nearly two hundred dead lakes have been identified in the Adirondacks alone. Most of the pollution responsible for the problem in this area seems to originate in the Ohio Valley and the Great Lakes area, and is then carried by prevailing winds into New England and southeastern Canada. In 1981 both countries entered into formal negotiations on ways to control the problem. Preliminary estimates are that it will cost Canada $500 million and the United States between $5 and $7 billion to halve the offending emissions.

Although it may be at its worst in the Northeast United States, acid rain and its consequences are by no means restricted to that area. There is evidence that it is on the rise in the West as well, even in remote areas far from major sources of industrial pollution. In 1980, William M. Lewis and Michael C. Grant of the University of Colorado at Boulder reported in *Science*[1] that the average pH of a series of remote streams on the western slope of the Rockies had dropped from 5.43 in 1975 to 4.63 in 1978, and that the drop was

---

releasing potentially hazardous substances like lead, cadmium, and asbestos into the water supply. As the pH drops below 5, natural aquatic organisms, especially fish, have a harder and harder time surviving in it.

continuing at a rapid rate. Lewis and Grant suggest that complex, poorly understood meteorological processes may be spreading pollutants from heavily industrialized areas, but there are other possible explanations for the appearance of hyperacidity in the Rockies, especially since it appears to be caused primarily by nitric rather than sulfuric acid. The sulfur emissions in the East are in large part due to the combustion of coal, which is not used as extensively in the West. However, as we mentioned in the previous chapter, nitrogen oxides are one of the major pollutants caused by automobiles, and they are among the most difficult air pollutants to control, whether they come from automobiles or from industries and power plants. It is very possible that the predominance of nitric acid in the Rockies is caused by the abundant presence of automobiles in the West, and by the relative absence of coal-burning power plants.

Acid rain does more than just kill lakes. It also corrodes buildings, damages agricultural crops, and possibly inhibits the growth of some types of trees, though it may help others. Furthermore, it may also play a role in the formation of methylmercury, a strong nervous system toxin that can occur in natural waters.

It seems likely that acid rain will increase in the 1980s. We will be burning more fossil fuels, especially coal, and the Reagan administration has already begun relaxing our Clean Air Act standards to encourage business. Some researchers expect to see 50 percent more sulfur dioxide in the air by 1985, and they all point to the total absence in this country of effective nitrogen oxide emission controls as a cause of concern. There are ways to counteract acidification of natural waters once it has occurred, such as by dumping large amounts of an alkaline substance like lime in them to counteract the acid, but these are costly and temporary. In the long run they will never be able to compete with the obvious advantages of effective prevention.

TOXIC WATER POLLUTANTS

Unlike conventional pollutants, toxic water pollutants tend to be those that are not easily biodegraded, are primarily a problem in highly industrialized countries, and for which we are only beginning to develop effective, economical control methods, Generally the kinds of health problems caused by toxic pollutants do not involve infectious diseases caused by living organisms like bacteria or viruses.

There are, of course, gray areas between "conventional" and

"toxic" pollutants. Nitrates, which we described above, are caused by animal wastes more often than by artificial fertilizers, but they typically only become a health problem in countries where such fertilizers are used heavily. Furthermore, the health disorder they can cause in humans is not an infectious one.

Establishing the true disease potential of most of the newer toxic industrial water pollutants can be an extremely difficult and controversial task. As one EPA official cogently remarked to us:

> There's a whole series of questions involved here. First, can we detect something in water? If we can, so what? Do we know that we can definitely establish a clear cause-and-effect relationship between it and disease? If so, are we talking about disease in rats, mice, and beagles or are we talking about clear evidence of disease in humans? If we're talking about humans are we extrapolating from animal studies or are we using good epidemiological data? Hopefully both. But it is very rare that we have good epidemiological data about the original exposure experience of our study population. And, then, finally, in the case of carcinogens, if we feel we have clearly established a specific cause-and-effect relationship between a specific substance and cancer in human beings, again so what? Does that mean that no matter what the risk level appears to be we have to try to completely eliminate that substance from our waters? Is that really feasible? Usually not, obviously. Or do we assume that it's reasonable to expect members of this society to accept some health risk in exchange for reasonably clean water at a reasonable cost? Ultimately what it seems to boil down to is how much are we really willing to pay, as a society, for protection from hazardous substances when we're not sure precisely how harmful they truly are and can't be sure until long after the damage has already been done? And to top it all off, we're not at all clear about who in this society should be making all these decisions about how much risk is acceptable.

It has become painfully clear in recent years that our technological ability to detect pollution in water and other substances far exceeds our ability to evaluate its health risk. We have the ability today to detect the presence of far more substances in a glass of ordinary tap water than we can even identify, let alone assess for their health effects. For example, in a recent analysis of water taken from a tap in Cincinnati, Ohio, over 700 separate compounds were found, of which only 460 or so could be identified. Of those identified, few had ever been reported before in raw or treated water supplies. Many are known or suspected carcinogens, though probably not at the extremely low concentrations reported in this study.[2] And even among those that we know or strongly suspect to be toxic, more often than not we have no true idea of the kinds of exposures to them that are clearly hazardous.

Despite these complexities there are substances whose toxicity is well established, whose current or potential prevalence in our water is widespread, and whose presence has until now eluded our ability to control them effectively. These of course deserve our unequivocal concern, though we may not yet know precisely what concentrations and exposures pose what kinds of health risks.

Mercury, our first example, is a natural mineral that has become a water pollutant in some parts of the world because it was found to be useful in a number of industrial processes and because it can be converted by industrial or natural aquatic processes into an extremely toxic organic form called methylmercury. Our second example, trihalomethanes (THMs), are ironically caused by the interaction of chlorine with natural organic substances in water. The most common THM, chloroform, is an animal carcinogen. Trichloroethylene (TCE), our third example, is a synthetic industrial solvent and one of the most ubiquitous groundwater pollutants found in this country to date.

MERCURY

Mercury is a very dense silvery-white metal that possesses the rare property (shared by only one other element, bromine) of being a liquid under normal conditions. Mercury is occasionally found in its pure form, but it is most commonly derived from cinnabar, a bright red mineral that is some 86 percent mercury.

Inorganic mercury has a long, colorful history as an occupational hazard. Like every other liquid it has a vapor pressure, which means that a small fraction of any amount of mercury exposed to air is constantly in the process of turning into a gas. Mercury fumes have been recognized as a poison since antiquity, plaguing such professions as mercury miners and smelters, gilders, and mirror makers. Mercury poisoning among hat makers, where it has traditionally been used to soften the fine hairs used in making felt, was the inspiration for the Mad Hatter in Lewis Carroll's *Alice's Adventures in Wonderland.*

A synthetic organic form of mercury has been used as a pesticide since the early nineteenth century. In 1914 the development of organic mercury fungicides led to a revolution in seed treatment methods that found its way to the United States some thirty years later. For many years organic mercury fungicides were used to treat the seeds of everything from oats to cherries here, and they are still used in a limited way for this same purpose today.

There are two clinically distinct but closely related forms of mercury poisoning. Historical mercury poisoning, caused by inhalation of mercury fumes, is almost exclusively a work site hazard. It is characterized by weight and appetite loss, insomnia, dermatological problems, mouth and gum sores, increased salivation, loss of teeth, halitosis, stomach pain, kidney problems, and a host of central nervous system disorders, including memory loss, decreased attention span, and diminished mental capacity. It can also cause a bizarre psychological disorder known as "erethism," from the Greek word for irritability. Erethism is described by two well-known industrial toxicologists, Alice Hamilton, M.D., and Harriet L. Hardy, M.D., in this way:

> There is a nervous timidity and shrinking from observation, a sense of discouragement, a loss of self-confidence, nervous fear of ridicule or criticism, fear of losing one's job, loss of joy in life. . . . Among the mercury miners and metallurgists in California, erethism or mercurialism is very familiar and the men recognize it in themselves as well as in others. They describe an increasing shyness, anxiety, embarrassment at being noticed, loss of self-confidence, or irritability which is very marked if the victim is spoken to suddenly or asked to do something unusual.[3]

Another classic sign of a nervous system affected by mercury is tremors. At first a poisoning victim's fingers, tongue, and eyelids shake rhythmically but almost imperceptibly, the tremor becoming more pronounced only if he or she is asked to perform a distinct action, such as touching his or her nose, while being watched. Periodically checking a mercury worker's handwriting is a traditional way of looking for early signs of mercury poisoning. As it becomes worse, it affects the muscles of the arms and legs, making walking, shaving, and other similar activities difficult or impossible. At this stage the regular, incessant fine tremors will also occasionally be interrupted by a sudden spasmodic jerk or twitch. Curiously enough, the tremors disappear when the victim sleeps and are mitigated by the performance of routine physical motions on the job.

Organic mercury poisoning can be caused by eating seeds treated with mercury fungicides, and in the 1950s and 1960s a series of health disasters involving mercury-treated seeds broke out in Iraq (1956 and 1961), Pakistan (1961), Guatemala (1965), and New Mexico (1969). In all these cases seeds treated with organic mercury fungicides and intended only for planting had either been baked into bread or fed to animals that were later slaughtered and eaten by the victims.

Organic mercury poisoning can also be caused by water pollution. These cases are rarer, but in the few incidents that have occurred, the health consequences have been grim.

THE DISEASE OF THE DANCING CATS

In 1956 an epidemic of organic mercury poisoning broke out in Minamata, a small town of fifty thousand residents located on the coast of Kyūshū, Japan's southernmost island. According to the authors of the best account of the incident,[4] W. Eugene and Aileen M. Smith, the disease had probably been smoldering in the Minamata area for decades, but had previously been confused with alcoholism or even the latter stages of syphilis. The process of correctly identifying it as the result of industrial pollution began when a five-year-old girl was admitted to a local hospital, unable to speak or walk properly, evidently suffering from a central-nervous-system disorder. Shortly afterward, the Smiths report, her younger sister was also admitted, with virtually identical signs. Then other members of the family were stricken. Soon neighbors were, too.

The illness at first defied diagnosis. Early guesses were that it might be an infectious disease, perhaps a form of polio. Japan had been struck by a virulent form of encephalitis in 1924. Spread by mosquitoes, it caused muscle spasms, mental confusion, and partial paralysis in arm and neck muscles. Health workers naturally tended to assume that the first cases of this new disease were an encephalitis recurrence. But as these possibilities were one by one ruled out by negative lab tests, medical suspicions began to turn toward the possibility of environmental poisons.

In Minamata in the mid-1950s the only significant industry in town was a large chemical plant owned and operated by the Chisso

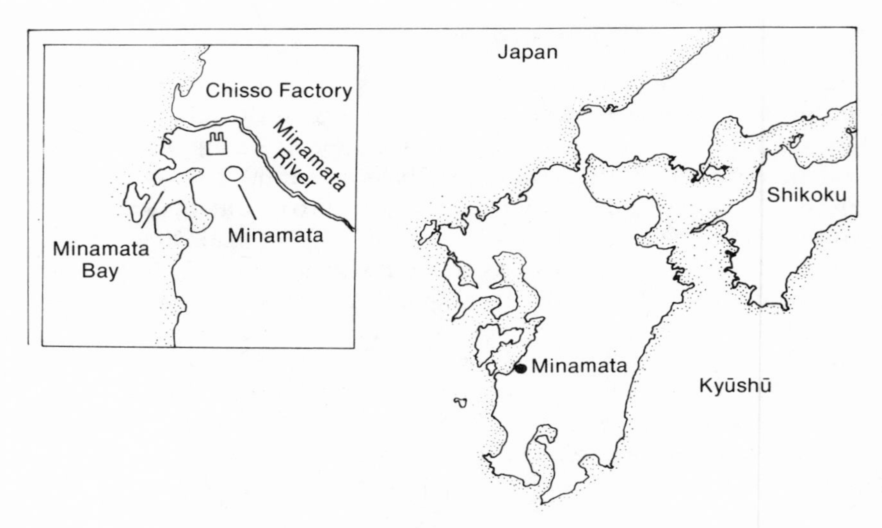

Figure 16

Corporation. The plant had been built in 1907. As the years passed, and Chisso became a major Japanese plastics and petrochemical company with a reputation for quality, it steadily expanded, bringing prosperity to the town. However, as the Smiths point out, by the mid-1920s the plant was already routinely dumping its wastes into Minamata Bay and just as routinely paying off the small suits brought against it by fishermen who claimed the wastes were ruining their livelihood by killing the fish and shellfish in the local waters.

There were early signs that something strange was happening in Minamata long before cases of human illness began to appear—birds fell out of the sky as though shot, cats whirled and leaped in the streets in crazy, often fatal convulsions—but no one paid much attention.

Although Chisso's wastes quickly rose to the top of the list of suspected causes of the strange new illness, which soon became known as "the disease of the dancing cats," no one had any idea which particular one of the sixty or so potential toxins found in the company's effluent might be at fault. As the possible toxins were tested, an indictment of methylmercury slowly emerged. Samples taken from the mud of Minamata Bay near Chisso's effluent outlet were found to be contaminated with over 2,000 ppm of mercury in some areas. By contrast, sediments taken from just outside the bay never exceeded 3.5 ppm of mercury. Fish and shellfish samples taken from the bay also contained high methylmercury levels, ranging from 10 to 40 ppm. The Smiths report that cats fed on fish and shellfish containing high methylmercury levels were found to have concentrated it at even higher average levels, 37 to 45 ppm, in their livers. Control groups of individuals from outside the Minamata area had at most only 2 to 3 ppm of mercury of tissues taken from any organ site. By contrast, autopsies of victims of the disorder in Minamata revealed levels as high as 70 ppm in liver tissues, 144 ppm in kidney tissues, and 24 ppm in the tissues of the brain, which is acutely sensitive to methylmercury.

Even residents of Minamata with no outward signs of the disorder had levels ranging from 100 to 190 ppm in hair samples; normal levels run about 8 to 9 ppm.

Until the early 1960s, it was widely assumed that elemental mercury emitted or dumped into the biosphere was not much of a hazard because it does not react easily with other substances, is quite stable, and above all, is only very slightly soluble in water. It seemed that industrialized nations could safely dump mercury into the world's waterways for centuries and all it would do would be to sink down to the sediments at the bottom. In healthy rivers and lakes the constant trickle of new layers of sediment caused by floods, soil erosion, and dying plants and animals seemed doubly to assure that mercury that found its way to the bottom of a body of water would soon be harmlessly buried under inches of dirt and decayed organic matter, safely isolated from the environment.

Then in the early 1960s Swedish environmental scientists began to wonder why it was that, although nearly all the industrial mercury wastes dumped into the waters of that country were in an inorganic, elemental form, over three-quarters of the mercury found in the tissues of the fish that inhabited these waters was of the more toxic alkyl form, methylmercury. By 1972 it had been repeatedly demonstrated in many parts of the world that microorganisms living in the bottom sediments of natural waterways had the power slowly to transform the relatively insoluble form of elemental mercury into the highly soluble and highly toxic form of methylmercury. Once it has been "biotransformed" in this way, methylmercury was easily absorbed by the tissues of fish in the same waters, and from there it easily passed into the tissues—especially those of the kidney, liver, and brain—of human and other predators that consumed fish contaminated with methylmercury. It has also been suggested that natural bacteria living in the intestines of rats and humans can cause this ominous conversion, biotransforming elemental mercury in foods or water into the methylated form:

Photo credit: Center for Creative Photography

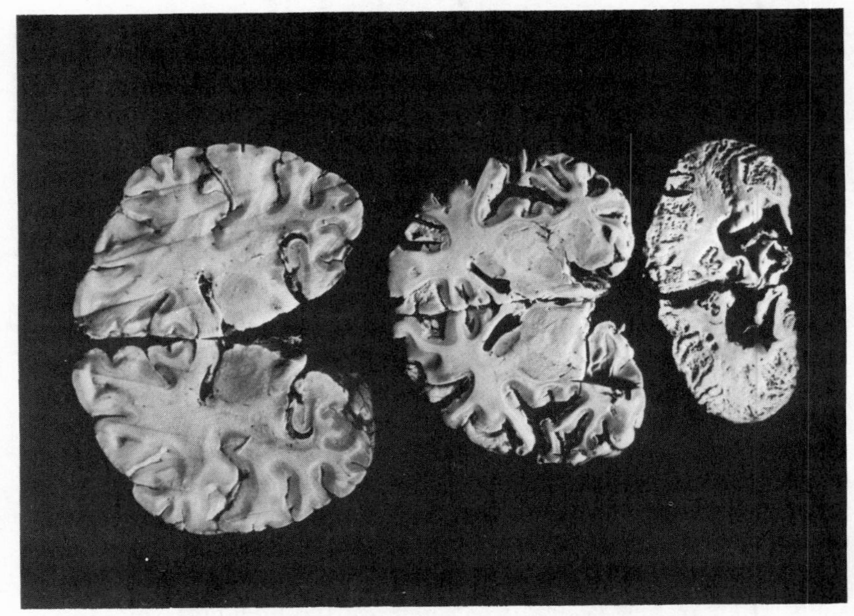

Brain Sections © The Minamata Project W. Eugene and Aileen Smith

Figure 17

> In an extension of the present work, we have shown that bacteria isolated from human feces can methylate mercuric chloride, which suggests that synthesis of methyl mercury compounds, from mercury present in food, can occur in the human intestinal tract, and that the present results from the rat can be extrapolated to the human situation.[5]

We also know now that other substances, including tin, chromium, selenium, arsenic, lead, platinum, thallium, palladium, gold, and sulfur are also subject to methylation by natural microorganisms, although we do not know how hazardous the infiltration of these methylated elements into the food chain is.[6]

The response of the Chisso Corporation to growing suspicion that its wastes were at fault was denial, misdirection, and even attempted intimidation of its growing chorus of accusers. Shortly after the 1956 outbreak, Chisso's own physician, Dr. Hajime Hosokawa, had quietly begun to test Chisso's wastes by feeding them to laboratory cats and observing the effects. On October 7, 1959, he found what he and everyone else was looking for. After acetaldehyde wastes containing mercury had been fed to cat #400, "it convulsed, salivated, and then suddenly whirled at great speed, crashing into laboratory walls."[7] Tragically, Dr. Hosokawa did not publish his findings until well over a decade later, in 1970, when he was on his deathbed. Fortunately they were still of great value at that late date, as evidence in the lawsuit that had finally been brought against Chisso by its victims the year before.

Chisso of course encouraged the view that natural processes rather than its plant were responsible for the methylation of the mercury found in Minamata Bay, but today the generally accepted view among scientists is that just the opposite was true: although bacteria in bottom sediments did cause some methylation, the wastes at Minamata already contained high levels of methylmercury when they were dumped into the bay. Prompt publication of Dr. Hosokawa's results with cat #400 would have helped resolve this ambiguity much sooner and would have helped save incalculable human suffering.

At the time, despite mounting evidence incriminating its effluent as the source of the problem, Chisso continued to deny all allegations, maintaining that insoluble elemental mercury could not make its way into water and thus the food chain of the bay.

In 1958 Chisso quietly moved its effluent outlet from the bay to a nearby river that also drained into the bay. Within months people living along the river had contracted Minamata disease. In 1959, still steadfastly denying any responsibility for the disease, Chisso with great fanfare announced that it was installing a "Cyclator," a revolutionary new type of waste water treatment unit that would remove any possible trace of harmful substances from the plant's effluent in the future. Within a year, according to the Smiths, Chisso's engineers knew that the Cyclator was not really as effective as the town had been led to believe, and they began to bypass it routinely. None of this was publicly disclosed until over a decade later. Mercury-laden wastes continued to pour into the bay.

There followed a lull that lasted for almost ten years. A strange stalemate, despite the regular diagnosis of new Minamata disease cases, prevailed until 1966, when an identical case of mercury poisoning broke out in the town of Niigata in central Japan, north of Tokyo. In this case the victims promptly sued the suspected source, the Showa Denko Company, and proved it guilty of dumping toxic substances into the Agano River. Their success ignited slumbering

Photo credit: Center for Creative Photography

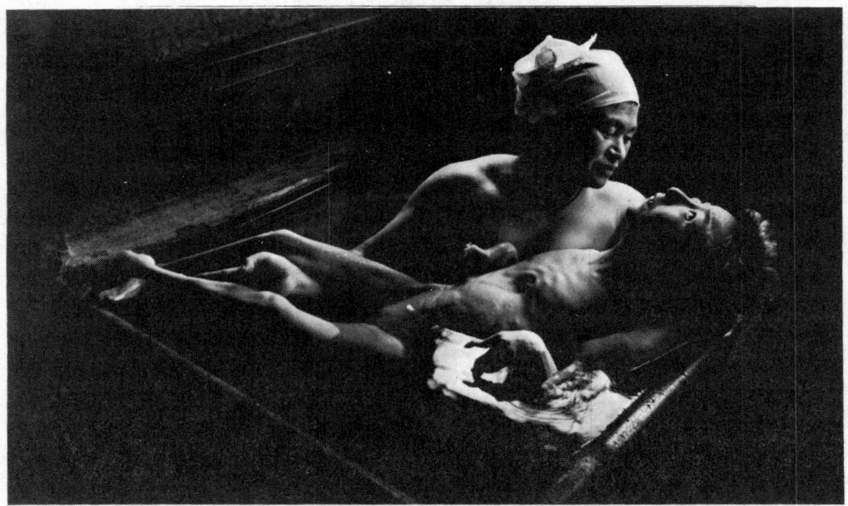

Tomoko in Bath © The Minamata Project W. Eugene and Aileen Smith

Figure 18

anger in Minamata, and three years later a suit was brought against Chisso. For the next four years, until the suit was finally settled, the incident was marked by increasingly militant demonstrations on the part of the victims and their supporters, including sit-ins, sleep-ins, hunger strikes, round-the-clock picketing of the company headquarters in Tokyo, and even occasional pitched battles with Chisso guards in the corridors of the corporate offices in Tokyo and in the plant at Minamata.

Not until 1968 did Chisso finally stop dumping mercury wastes into Minamata Bay, and only then because the practice of using mercury catalysts in making acetaldehyde had become obsolete, not because of the health problems it had been causing.

On March 20, 1973, a Japanese court decided the case in favor of plaintiffs and ordered Chisso to pay each one of them, including the heirs of those who had died, from sixty thousand to sixty-eight thousand dollars, depending on the severity of their condition. One week later, the Smiths report, representatives of the victims negotiated a second agreement in which Chisso agreed to bear the costs of medical care and living expenses for each Minamata victim. The president of the company publicly apologized for Chisso's negligence.

As of the end of 1979, 1,401 individuals had been certified as victims of Minamata disease. Of this number 353 had died. Chisso had paid out over $200 million in damages and had been economically crippled in the process. In recent years the local prefectural and national Japanese governments have had to step in to back up payments to Minamata victims because of Chisso's increasing inability to handle the financial burden of the payments.

COULD IT HAPPEN HERE?

As early as 1956 the FDA had ordered a halt to the use of organic mercury as a slimicide in the manufacture of paper and cardboard because of fears that it might contaminate foods stored in containers made of these materials. In 1971 the USDA canceled the registrations of many of the organic mercury formulations widely used before then as fungicides. Since then the use of mercury of all types in the United States has been steadily declining. Today the major users of elemental mercury in this country are probably dentists, who still use mercury in the process of making fillings; most of us carry some mercury around in our teeth.

There are, however, other sources of mercury emission into the environment at large that may pose an even greater water pollution hazard. In September 1970 health officials declared the North Fork of the Holston River, near Saltville, Virginia, off limits to sustenance fishing because mercury levels in fish taken from the river exceeded the FDA standard of 0.5 ppm. The source of the mercury was a chemical plant nearby, built in 1895, most recently owned and operated by the Olin Corporation, and defunct since the early 1970s. From about 1951 until it was closed the plant produced chlorine using mercury in an electrolytic process. After use the mercury was passed along with waste water into large disposal ponds on the site, and some of it settled into the soil beneath the chlorine plant or the ponds, then gradually leached into the river. Today experts calculate that there are still some hundred thousand kilograms of mercury remaining in the soil beneath the old chlorine plant and almost forty thousand kilograms in the soil beneath Pond 5, where most of the mercury-contaminated waste water was channeled. They also believe that the mercury sediments in the riverbed are releasing only about twenty-two kilograms of mercury a year into the river.

After 1970 company and local officials kept a close watch on the mercury levels at the site, hoping that natural dispersal would remove the problem. But in the late 1970s mercury levels in fish taken from the Holston rose and it became apparent that something would have to be done to stop the steady migration of old mercury residues into the river.

Removing the contaminated soils completely would have cost some $25–30 million and was considered too expensive. Instead, the company and the state have elected to take simpler steps. Riverbanks near the site have been regraded to reduce the erosion that was causing much of the deposition of new mercury wastes into the river. Old drainpipes at the site have been removed or plugged. New drainage diversion channels have been dug, and old ones widened. Sand and topsoil have been applied over polluted areas. The surface of Pond 5 may be sealed. But it's still too early to know how effective these steps will be. Fortunately, because of Minamata and other incidents, concern over mercury poisoning led to control measures soon enough so that no human illness has occurred because of the Holston pollution problem.

Another potential and less controllable source of mercury contamination of the environment, including waterways, is the combustion of fossil fuels, particularly coal. Mercury levels in coal are higher, on the average, than in any other fossil fuel. Crude oil, for example, contains anywhere from 1.9 to 21 ppm mercury. Coal can hold from as little as 0.5 ppm to as much as 46 ppm elemental mercury.

In 1971 researcher Oiva I. Joensuu reported in *Science* on the results of tests in which thirty-six different types of American coal were analyzed for mercury content.[8] Based on his results Joensuu estimated that a yearly global coal consumption of 27 billion metric tons would release 2,700 tons of mercury into the environment. We currently burn about 680 million metric tons in the United States. How much of this might find its way from the air into waterways where it may be subject to significant methylation, and how much of a hazard natural methylation might pose, is a matter of guesswork.

When coal is burned in a power plant's combustion chamber, the mercury in it vaporizes and travels up the stack. The use of devices to precipitate it will capture some amount, depending on the efficiency of the system, and the use of scrubbers, misters, or other additional cooling mechanisms in the flue will reduce the emissions levels still further. But inevitably some percentage of mercury vapors and particulates will still escape into the ambient air.

Once mercury is emitted into the ambient air, either as a gas or as liquid droplets, its movement depends, like that of any other emission, on such factors as wind, rain, and dust conditions. Because mercury has a very strong tendency toward adsorption, it readily attaches itself to other forms of particulate matter, especially in air that is heavily laden with dust. As these particles grow heavier, they naturally tend to sink to earth, or to be washed out by rain or snow. Eventually they are redeposited on plants, buildings, trees, the soil, or on the surface of bodies of water. Much of the mercury deposited on land is washed off by subsequent precipitation into natural waterways, where it adds to levels of mercury deposited by direct air-to-water surface transfer, or from industrial dumping.

Elemental mercury is not biodegradable. It sinks to the bottom of bodies of water and collects there, available for biotransformation by methylating microorganisms.

We do know that methylation of mercury does take place in numerous places throughout the United States, though at concentrations that appear to be well below those that could pose a threat to human health.[9]

We also know, however, that the sulfur and nitrogen oxides that are also emitted by the combustion of coal and other fossil fuels and that are largely responsible for acid rain, enhance the methylation process:

> Another metal, mercury, may produce lethal effects as a result of acid rain. As lakes acidify, mercury goes into solution as highly toxic organic methyl mercury. A clear correlation exists between the acid level of the lake and the mercury levels of its fish; the more acid added to the lake's waters, the more mercury in the fish, up to the point where the fish may begin to die. Those fish that survive present a clear danger to man—mercury, at even extremely low levels, can be lethal once it is introduced into the food chain and consumed by humans.[10]

CHLORINATION AND CANCER

Epidemiologists have known for some time that there is a relationship between surface sources of drinking water and elevated cancer rates, but the reasons for this connection remain somewhat mysterious. One common theory is that surface waters are more vulnerable to pollution by the countless toxic substances that industrial societies release into the air, into the soil, and dump directly into the water. If their theory is true, then the increasing industrial contamination of groundwater aquifers that we are currently witnessing should erase the difference in cancer rates: either groundwaters will become just as contaminated as surface waters, in which case they will cause just as much cancer, or they will become *more* contaminated than surface sources, in which case those using them for drinking water will be forced to switch to other sources, most of them surface waters, and the basis for the difference will then also be eliminated.

One of the earliest studies linking surface water, organic contaminants, and increased cancer rates appeared in August 1977. Prompted by two previous studies in 1975 in which EPA researchers had determined that the problem of surface water contamination by a number of hazardous synthetic organic chemicals was widespread, the authors of this study decided to compare cancer rates between the forty-six Ohio counties served primarily by surface water sources and the forty-two served primarily by groundwater sources.

Their results showed that white males living in areas served primarily by surface-water drinking supplies had a 5 percent higher death rate from all cancer types than males living in groundwater areas, including a 10 percent higher rate for stomach and an 18 percent higher rate for bladder cancers. White females in surface-water counties exhibited a significantly higher rate of stomach cancer deaths as well.

As suggestive as these results were, the study was only a preliminary step, one that in no way specifically incriminated chlorination or its byproducts. As the authors themselves pointed out:

> While the study is only a first step in assessing the health affects [sic] of these chemicals and by no means establishes a causative factor, it suggests that organic chemicals or other substances in the water supply may relate to increased cancer mortality in communities served by surface water supplies. It further indicates the need for more specific epidemiological studies using more direct water quality assessments.[11]

Recently, and ironically, suspicion about one possible cause of this relationship has fallen on a group of chlorination byproducts known as trihalomethanes (THMs). Trihalomethanes—one carbon atom plus one hydrogen atom and three halogen atoms—are part of a larger chemical family known as halogenated hydrocarbons, a group of clear, volatile liquids widely used in industry as solvents, aerosol propellants, coolants in refrigeration systems, and in the manufacture of plastics, drugs, and rubber. In the workplace the most significant health hazard posed by halogenated hydrocarbons is inhalation of their vapors, which can cause liver and kidney damage and depression of the central nervous system.

Trihalomethanes are formed when chlorine or, more rarely, bromine or iodine react with substances known as trihalomethane "precursors," which are given off by decaying organic matter. Peat, for example, as it decomposes in water is broken down to humic and fulvic acids, both of which are THM precursors.

Trihalomethanes are carcinogenic in laboratory animals. They are found only in treated water, and traditionally this has meant surface water. Some groundwater used to provide community drinking water is routinely chlorinated, but most groundwater is used by individual well owners and is not treated at all. Thus the difference between surface water and groundwater cancer rates may also be, in part, a difference between chlorinated and unchlorinated water.

Of a total of ten possible trihalomethanes, four are found commonly. Because they are formed, measured, and managed in very similar ways, and because their toxicological properties are quite similar, they are usually referred to collectively as total trihalomethanes, or TTHMs. Chloroform ($CHCl_3$), used as a surgical anesthetic until it was found in the 1950s to cause such liver diseases as hepatitis in patients, is the most common trihalomethane found in treated drinking water. The other three are bromochloromethane, dibromochloromethane, and tribromochloromethane, also known as bromoform.

In 1975 an EPA survey of eighty water supply systems around the country detected chloroform levels ranging from a low of 0.4 μg/liter in Albuquerque to a high of 311 μg/liter in Miami, Florida. Eight places across the nation were found to have TTHM levels exceeding 100 μg/liter, which is now the Maximum Contaminant Level (MCL) allowed by law for TTHMs. Follow-up surveys in 1976 and 1977 found even more locations—a total of thirty-one—with TTHM levels that exceeded 100 μg/liter.

By 1980 it appeared that the link between TTHMs and cancer had become even stronger. Further epidemiological assessments of thousands of cancer fatalities in New York, Illinois, Louisiana, North Carolina, and Wisconsin conducted by the federal Council on Environmental Quality (CEQ) reinforced the relationship between increased risks of rectal, bladder, and colon cancer and high levels of TTHMs in the water, specifically the chloroform associated with chlorination.

The evidence found by the CEQ seemed so convincing that it

was moved to appeal to the EPA to lower the proposed MCL for TTHMs, on the grounds that 100 μg/liter did not provide adequate public protection. Despite this controversy the 100 μg/liter MCL still stands, and as of this writing was scheduled to take effect in November 1981 for cities with populations over seventy-five thousand and two years later for smaller communities.

Nonetheless, the TTHM issue is still highly controversial. A number of water quality experts are not at all convinced that the link between TTHMs and cancer has been conclusively demonstrated. For example, Robert W. Tuthill and Gary S. Moore of the University of Massachusetts compared chlorine levels, THM levels, and cancer mortality rates for nearly every community in Massachusetts with more than ten thousand people, and found significant associations, as have many other investigators, between chlorination and stomach and rectal cancer rates. However, the association was greatly weakened when data on ethnic background and place of birth, both known to be related to certain types of cancer, were factored into the statistical calculations. Tuthill and Moore concluded that:

> The results of this investigation indicate that trihalomethanes and other chlorination byproducts are not statistically significantly associated with cancer in the Massachusetts communities studied. The practice of chlorination should not be phased out in preference to another method of disinfection until there is stronger epidemiological and clinical evidence to support the charge that it produces serious health risks, and until the potential health effects of substitute disinfectants are carefully explored.[12]

The controversy over THMs may soon become moot, because it appears that even if they do pose a problem, we will be able to control it fairly easily. There are a number of effective, economical ways to minimize THM formation in water supplies. One is to reduce the amount of precursor in the water so that there will be less of it for chlorine to react with. Another is to use something other than chlorine as the disinfectant. Possible alternatives include chlorine dioxide, chloramines, and ozone, none of which will react with precursor matter to form THMs. Industrial-strength granular activated carbon (GAC) filters will also remove THMs from water.

Naturally, some of these alternatives have their own drawbacks. Chlorine dioxide, for example, though it is an extremely effective disinfectant, especially for removing viruses from water, is highly explosive and extremely toxic. As one water quality expert told us, "You've got to make it at the treatment plant—you can't transport it or store it in tanks. A standard chlorination system is dangerous enough: If you smell a leak, you can get out of there. But with chlorine dioxide, by the time you can smell it you're dead."

Ozone, used along with chlorine in parts of Europe and in a few experimental water treatment plants in the United States, is difficult to use because it disappears soon after it enters the water, which is then vulnerable to recontamination. It is also more expensive than chlorine and requires more complicated application equipment.

Currently, the most promising THM removal method is something called "air stripping," or "gas stripping," which basically means nothing more complex than blowing air through the water. Because THMs are volatile, meaning that they easily turn into a gas, this procedure causes them to leave the water and mix with the air. So far, the process looks simple and inexpensive.

TCE IN THE GROUNDWATER

One of the most urgent water pollution problems we face today—pollution of groundwater resources by synthetic organic substances—has ironically been aggravated by our own efforts to clean up other parts of the environment. During the 1970s industries found themselves in a squeeze: Concern about the quality of the air, the oceans, and our surface waters brought about a series of laws prohibiting or severely restricting uncontrolled emissions and discharges of wastes into the biosphere. One by one traditional ways of getting rid of wastes were outlawed. Instead of freely releasing potential pollutants into the environment, industry was more and more often forced to pretreat them heavily, and in so doing many substances once casually dumped or vented into the air were captured and contained in drums, leading to a second problem: What to do with the accumulated drums? Usually industries did the obvious thing—hired someone to haul them away. But all too often this just displaced the pollution problem rather than solving it, since the drums frequently wound up sitting in an open field or shallow trench, or their contents were poured out into open ponds, abandoned mine shafts, or right onto the ground. When ponds or abandoned drums began to leak into the soil below, as most of them eventually did, there was almost never any barrier between them and local groundwater resources. As a result we are finding today that aquifers all across the country are either already highly contaminated or are in imminent danger of contamination by a variety of toxic compounds.

The EPA estimates that some 50 million metric tons of hazardous industrial wastes are disposed of each year in landfills or waste lagoons that pose a threat to nearby groundwater resources because of improper construction. A 1979 survey by the EPA counted more than 130,000 waste lagoons, as many as three-quarters of which lacked the basic components—such as clay linings—needed to protect groundwater supplies from contamination. In the case of some toxic chemicals, like TCE (trichloroethylene), even clay linings wouldn't be of much use, since sufficiently high TCE concentrations will cause clay to break down. Furthermore, it doesn't take much leakage of the more common synthetic organic compounds to cause a high level of water pollution: One-half kilogram of TCE will contaminate 100 million liters of groundwater to a concentration of 5 μg/liter, higher than the emergency standard recently adopted in many areas of the country.

As a result of these and other improper disposal practices, some dating back to the 1950s, significant contamination of aquifers by chemicals like TCE has already occurred in many parts of the country—Arizona, California, Connecticut, Florida, Massachusetts, Michigan, New York, and other areas. According to some eight thousand tests conducted by state water quality agencies all across the country and reported in March 1980, TCE is probably the most common groundwater pollutant in the nation today.

TCE is a clear, sweet-smelling liquid that readily vaporizes into a flammable gas. It is part of the chemical family known as aliphatic halogenated hydrocarbons, closely related to THMs, described above. Members of this family have found a multitude of industrial uses over past decades as solvents, metal degreasers, coolants, pesticides, and cleaning agents. They have also found applications as ingredients in adhesives, lacquers, varnishes, paints, and inks. At one time TCE was even used in the decaffeination of coffee beans, and has also occasionally been used to help keep septic tank lines free of obstructions. One of its earliest uses was as a childbirth anesthetic, in a form called Trilene. That use was discontinued in the 1950s when TCE's properties as a kidney and liver toxin were learned, and because it was found to be highly flammable at about that time.

As a chemical class the aliphatic halogenated hydrocarbons share many of the toxic effects of other industrial solvents. They can cause skin and mucous membrane irritation, toxic hepatitis, and kidney disease. Chronic skin contact with water containing TCE or other similar solvents can cause irritation and rash. Ingestion of substances containing high concentrations of TCE has been known to cause respiratory failure and cardiac arrest. The body naturally converts TCE into chloral hydrate, a prescription sedative.

Many aliphatic halogenated hydrocarbons are animal carcinogens. Some, such as vinyl chloride, are also clearly responsible for certain types of cancer in humans, particularly in occupational settings. The current scientific view is that TCE is a liver carcinogen in mice at high doses. However, we do not yet know enough about TCE's carcinogenic properties to predict accurately the effects of chronic ingestion of small amounts of TCE in drinking water.

In one well-known pollution case in Jackson Township, New Jersey, a dump licensed to handle only municipal garbage mistakenly also accepted loads of toxic chemical wastes from local industries. The dump was situated directly over the Cohansey aquifer, from which hundreds of local residents drew their household water. By 1978 a rising chorus of health complaints from local residents, many concerning serious disorders, prompted New Jersey health officials to evaluate samples of water taken from wells in the township area. The results showed potentially harmful levels of a number of toxic industrial chemicals, including benzene, acetone, chloroform, methyl chloride, toluene, and TCE. Today, about one hundred of the wells that once served families in the township are closed and their water is now brought in by truck.

In late 1979 a routine water analysis conducted by a private firm in Southern California turned up surprisingly high levels of TCE in samples drawn from a local well. A subsequent testing program by California state health officials of some 431 wells in the San Gabriel Valley area disclosed unsafe levels of TCE in 59 of them, ranging as high as 770 μg/liter in one case. No one is quite sure how these wells became contaminated with TCE, but local officials surmise that industries in the area some fifteen to thirty years ago may have disposed of waste TCE from degreasing processes by simply dumping it on the ground. Unfortunately, the valley floor in that area is composed of an alluvial gravel, the bed of the prehistoric river that formed the valley, which is even less of a barrier to chemicals like TCE than the soil is in most other places.

The significance of a level like 770 μg/liter TCE becomes clearer in light of the 1977 estimate by the National Academy of Sciences (NAS)[13] that a population drinking a typical two liters of water per day for an average life span of seventy years will experience one additional case of cancer per million people if the water they drink is contaminated with as little as 4.5 μg/liter of TCE. More recently the EPA in conjunction with the NAS and other scientific organizations have halved that level to 2.7 μg/liter of TCE.[14] Fortunately, the vast majority of contaminated wells in the San Gabriel Valley contained levels far below the maximum 770 μg/liter found in one sample. Fewer than ten contained TCE levels above 100 μg/liter, and most were well below 30 μg/liter. An extensive tap-water testing program in the area revealed only two out of eighty-one samples with TCE levels above 5 μg/liter.

In 1980 a process of setting a national water standard for TCE and a group of similar substances was just beginning. Most water quality specialists expected it to be completed in early 1982. Unfortunately, under the Reagan administration the EPA has postponed hearings that were scheduled to be held all across the country as part of this process, and there is no indication whatsoever that they will ever be rescheduled.

In the meantime, at least one state, California, has adopted the EPA's recommended "action level" of 4.5 μg/liter (rounded off to 5 μg/liter in California) and has advised water suppliers in the San Gabriel Valley not to use water for human consumption if TCE levels exceed this concentration. All sixty-odd wells containing TCE concentrations above this level in the San Gabriel Valley were promptly closed to human consumption. Another 372 wells were tested and found to contain no TCE or TCE concentrations well below the action level. However, if the standard that is eventually adopted is as low as 2.7 μg/liter, it will mean that, as in Jackson Township, more wells in the San Gabriel Valley will be closed and some residents will have to look elsewhere for their water supplies.

THE 1970s: A DECADE OF CLEANUP LEGISLATION

Legal efforts to limit the water pollution caused by cities and industry go back decades in some states, but in the 1950s and 1960s it became clear that a piecemeal approach to water pollution control on a state-by-state basis was not working. As one area toughened its water quality standards, all too often the reaction among industrial polluters was simply to pick up and move to a state with easier standards. And since pollution does not honor abstract political boundaries like state lines, this often meant that even the "tough" states still had to contend with pollution spilling over from neighboring "easy" states.

It also became clear that traditional pollution control measures were just not working anymore in some states. In 1970 an infamous incident occurred in which Ohio's Cuyahoga River, which winds through the steelmaking region around Cleveland, actually caught fire due to the oil, grease, and chemicals accumulated in it. Tests of its quality at the time revealed that its bacteria count in places was equivalent to that of raw sewage. At about the same time, a federal sampling of nearly one thousand water supply systems in the United States serving some 18 million individuals disclosed that 13 percent of them were drinking water of questionable quality and some three hundred thousand were drinking water that was so polluted it was almost certainly hazardous.

For these reasons and others the 1970s became a decade of new water pollution legislation, particularly at the national level in the form of broad, ambitious federal laws that set up programs administered by the EPA and were aimed at eliminating pollution caused by cities and industry as well as protecting the consumer by identifying pollutants that pose clear health hazards and establishing maximum allowable concentrations for their presence in drinking water.

THE CLEAN WATER ACT

The centerpiece of our national campaign against water pollution in the last decade is Public Law 92-500, also known as the Federal Water Pollution Control Act Amendments of 1972 and 1977.* The very first sentence of the act announces the most comprehensive water pollution control program ever envisioned: "The objective of this Act is to restore and maintain the chemical, physical, and biological integrity of the Nation's waters." PL 92-500 goes on to list six deceptively simple goals and policies that its framers felt would help achieve this objective, including eliminating the discharge of all pollutants into navigable† waters by 1985, prohibiting "toxic pollutants in toxic amounts," and providing federal grants for the construction of municipal sewage systems, an area that the Reagan ax is already chopping at.

The act is subdivided into five major parts, or "titles." Title I

*The title was retroactively shortened to the Clean Water Act in 1977.

†This word, which seems unrelated to issues of pollution, is a vestige of the old River and Harbors Act of 1899, which authorized the federal government to protect routes of waterborne commerce. In the early 1970s this precedent was broadened by federal courts to include federal regulation of water pollution.

covers research, Title II outlines the granting system for community sewage plants, Title III describes the standards and enforcement system used to insure that pollutant discharges are brought under control, Title IV outlines the permit system used to identify and monitor the fifteen thousand communities and forty thousand industries licensed to emit specific concentrations of pollutants into the waterways, and Title V covers general administrative matters, including such topics as the EPA's role in water pollution emergencies, and the right of citizens to sue government agencies or individuals for failure to protect water quality. In short, the Clean Water Act makes water pollution illegal and establishes the legal and financial tools needed to phase it out by the mid-1980s. A more ambitious environmental cleanup program has never been seen before in our nation's history.

The provisions of the Clean Water Act that have had the most immediate and far-reaching effect on most of our lives are probably the establishment of a timetable under Title III that requires the implementation of progressively more sophisticated pollution control technology by specific deadlines and the corresponding authorization under Title II of substantial grants for the design and construction of better public sewage treatment facilities. There are a few government programs that help finance water treatment systems for businesses, especially small businesses, but in general the act follows the point of view that taxpayers should not be expected to subsidize the private, profit-making sector, and on the whole industry has agreed with this philosophy.

The first step in the process of eliminating water pollution was the logical one of inventorying all readily identifiable water polluters, private and public, recording what it is they discharge, and issuing them a permit that allows them to continue to discharge only if they keep the concentration of a series of different pollutants within safe levels. Such a system, called the National Pollution Discharge Elimination System (NPDES) and described in Title IV of the Clean Water Act, has been in operation now for about a decade. It is, fundamentally, a licensing system.

Unfortunately, not all sources of water pollution are readily identifiable city sewer systems, industrial plants, or specific irrigation drainage ditches. To use the phraseology of the Clean Water Act, there is a large difference between "point" and "nonpoint" water pollution sources. Point sources are those that can be easily detected and are associated with a specific facility, person, business, community installation, or other legally liable entity. As the law puts it, a point source is "any discernible, confined and discrete conveyance, including but not limited to any pipe, ditch, channel, tunnel, conduit, well, discrete fissure, container, rolling stock, concentrated animal feeding operation, or vessel or other floating craft."[15] In other words, if it's a point source, you know where the pollution is coming from and who is responsible for it, so you can issue an NPDES permit, which can then be revoked if the discharger doesn't comply with water quality standards.

But there are also plenty of sources of water pollution that don't come from a single, discrete, easily identifiable "point" source; these the Clean Water Act labels "nonpoint" sources. They include such things as the runoff in cities from water that flows into the storm drains, carrying with it all the old oil, grease, leftover antifreeze, and other assorted urban flotsam and jetsam that builds up on the streets and sidewalks between rains. In some cities, particularly older cities, the storm drains and sewers are combined in one system, and a good storm can overload it to the point that either sewage backs up or the treatment plant has to let it flow through untreated.

Other kinds of nonpoint water pollution sources include runoff from agricultural areas, containing substances like pesticides and bacteria from manure, runoff from logging and construction sites, containing lots of silt and sediment, as well as oil and grease and frequently some chemicals, acidic wastes from active or abandoned mine shafts, and seepage from ruptured chemical waste drums stored on or below the earth's surface.

On a larger scale, one that is not mentioned in the act, "nonpoint" pollution of surface waters takes place constantly via deposition from polluted air, either directly to the water surface as pollutant particles grow heavier and sink back to earth, or through precipitation.

Title III sets up the system that establishes and enforces standards intended to regulate the kinds and amounts of pollutants that can be allowed in treated water that is returned to the environment. In order to reach the national goal of completely eliminating the discharge of pollutants into navigable waters by 1985, the act sets deadlines for the gradual phase-in of steadily more effective waste-treatment methods by municipalities and industries.

Title II of the act sets up the legal framework for federal grants to communities or states that want or need to construct new waste water facilities or upgrade their old systems. For conventional systems the act authorizes the EPA to pay up to 75 percent of the costs of their planning and installation. Under the Reagan administration the amount of federal money available for these purposes has been cut drastically.

A strong incentive to seek out and support "alternative" and "innovative" waste water treatment methods was also built into the act. The difference between an alternative and an innovative treatment system is that the former have been demonstrated and are known to work, though they may not have found widespread application in this country for a variety of reasons. Examples of alternative approaches currently in operation or being planned throughout the nation include systems that use waste water to raise fish or compostable plants, and to irrigate crops, parks, and golf courses, that use wastes to fuel incinerators, and that dry or compost sludge for use as a fertilizer. Innovative systems, on the other hand, are those that, however sound they may be on paper, have never been tried before, and are radical departures from past experience.

For alternative and innovative systems the EPA is authorized to provide 85 percent of the cost of design and construction, even if their projected total life-cycle costs are as much as 15 percent higher than those for a conventional system. In addition the EPA is allowed to offer what is in effect a fail-safe guarantee: If for some reason an alternative or innovative system doesn't work out once it is in place, whether because it is too expensive, wastes energy, or just doesn't function the way it is supposed to, the EPA will pay 100 percent of the costs of replacing it with something that does work. Finally, any applicant for federal grants under Title II of the Clean Water Act must prove to the EPA's satisfaction that it has carefully considered appropriate alternative and innovative systems.

In fiscal year 1981 the EPA was authorized to spend an additional $3.2 billion on publicly owned treatment works before President Reagan's cuts reduced this amount by nearly one-third. Since 1973 about $25 billion has been spent on construction grants under this act, making it the largest program administered by the EPA, and one of the largest public works programs in the country's history.

The Clean Water Act sets stiff penalties for violators of its regulations—fines of from twenty-five hundred to twenty-five thousand dollars per day and one year in jail for the first conviction, double that—as much as fifty thousand dollars per day and two years in jail—for the second. Falsifying water quality records or tampering with monitoring devices can earn a violator up to ten thousand dollars in fines and a six-month jail sentence.

The CWA also empowers the EPA to respond quickly to a water pollution emergency, such as an oil spill. A $10 million contingency fund is called for in the act to subsidize research assistance, abatement and administrative activities, and legal support for subsequent litigation.

THE SAFE DRINKING WATER ACT

Even if the overall goal of the Clean Water Act—pristine waters again by 1985—could be achieved, we would still have to face the fact that some water pollution is inevitable. For example, as in the case of acid rain, the interdependence of the biosphere means that substances that pollute the air or the earth will in all likelihood also pollute the water. As long as our culture releases *any* pollutants from any source into any part of the biosphere, some of them will find their way into our water supplies. Moreover, nature itself "pollutes" through sources like volcanoes, mineral deposits, decomposing plant and animal matter, and forest fires. Even if we were somehow miraculously able to bring our own pollution to a sudden halt, we would still have these natural pollutants to worry about.

The essence of Public Law 93-523, the Safe Drinking Water Act, is the establishment of a system of identifying significant water pollutants and setting limits, called Maximum Contaminant Levels (MCLs), on their concentration in drinking-water supplies. Anyone who supplies drinking water to the public must comply with these standards, and that includes not only the 250,000 community water treatment plants around the country but also the estimated 200,000 or so restaurants, gas stations, bottled water companies, and so forth that also provide drinking water. Suppliers must not only regularly test their drinking water to make sure that no MCLs are exceeded, but also, when a pollutant level does rise above the MCL, must notify the public and advise it on steps being taken to avoid health hazards caused by the pollution. For a water treatment plant in a typical community, this usually means placing notices in local news media.

In effect, the act guarantees everyone the right to know where his or her drinking water comes from, how it is tested and treated to make sure that it is safe to drink, and whether any levels of potentially hazardous substances are exceeded. The law also guarantees that anyone who feels that his or her rights under this law are being abridged can sue the supplier, the state, the EPA, or all three. Violators of the act's regulations face fines in most cases of up to five thousand dollars per day.

Most water treatment systems were required to be in compliance with the law as of June 24, 1977. As new MCLs are passed, one year's grace period is typically allowed before drinking water suppliers must comply with it. Some smaller systems were granted exemptions until January 1981 to meet Safe Drinking Water Act standards, or even January 1983 if they were planning to become part of a larger regional water treatment system, but no exemptions or variances at all are permitted if local, state, or federal authorities feel that doing so would threaten public health.

States may set MCLs for substances they feel are of particular concern in their area, or they may set MCLs for substances specified in the law that are tougher than those established by the EPA, but they may not set MCLs that are less stringent than the federal standards.

As of mid-1981 these MCLs were in force:

Inorganic chemicals	0.05 mg/L
Arsenic	0.05 mg/L
Barium	1 mg/L
Cadmium	0.010 mg/L
Chromium	0.05 mg/L
Lead	0.05 mg/L
Mercury	0.002 mg/L
Nitrate	10 mg/L
Selenium	0.01 mg/L
Silver	0.05 mg/L
Fluoride*	1.4–2.4 mg/L
Organic chemical turbidity	1 tu up to 5 tu†
Coliform bacteria‡	1/100 ml
Endrin	0.002 mg/L

*Chronic consumption of excessive amounts of fluoride can cause fluorosis, an abnormal bone condition in which bone matter becomes simultaneously denser but weaker, and thus is much more subject to fractures. Fortunately this is only a problem at relatively high exposures, e.g., above 20 ppm, which occur only very rarely in nature in a few locations, primarily in parts of the Middle East, India, and Africa. As far as anyone has yet been able to determine, fluoride ingestion at the levels used to control tooth decay, i.e., 1 ppm, is completely harmless.

†The symbol "tu" stands for "turbidity units." Turbidity, the amount of suspended matter in water, is an important indicator of water quality because it may interfere with the chlorination process. Viruses, in particular, may be shielded from disinfection by high levels of turbidity. Turbidity is calculated by directing a beam of light at a sample of water and measuring the amount reflected.

‡The coliform test is a count of total colonies of bacteria developing from measured portions (1 ml) of water tested, planted in petri dishes with agar culture, incubated for twenty-four hours at 35°C or forty-eight hours at 20°C.

Lindane	0.004 mg/L
Methoxychlor	0.1 mg/L
Toxaphene	0.005 mg/L
2,4 D	0.1 mg/L
2,4,5 TP Silvex	0.01 mg/L
Radionuclides	
Radium 226 and 228	5 pCi/L*
Gross alpha particle activity	15 pCi/L
Gross beta particle activity	4 mrem/year
TTHMs	0.1 mg/L

During 1981 the EPA was considering establishing new MCLs for a number of other contaminants, primarily uranium and a group of synthetic volatile organic compounds (VOC), including:

Trichloroethylene
Tetrachloroethylene
Carbon tetrachloride
1,1,1-Trichloroethane
1,2-Dichloroethane
Vinyl chloride
Dichloroethylene
Benzene and chlorinated benzene compounds
Methylene chloride

WHERE ARE WE NOW?

What has been the effect of the past decade of intensified national concern over water pollution? Have the laws, the massive grant programs, the tougher new effluent standards made a difference? Is our water getting cleaner? Is it more fishable, swimmable, drinkable?

Not surprisingly, the answer is yes—and no. In some parts of the country the water is much cleaner, much healthier. In other areas we are only just beginning to understand the true dimensions of the pollution problems we face, and we have no good ideas yet on how to handle them. In most areas of the country we're somewhere in the middle: progress has been made, but the real job still lies ahead—what's been accomplished so far is at best still only a good start.

The Clean Water Act requires each state periodically to submit

*pCi/L = picocuries per liter. (We discuss radioactivity further in Chapter 8.)

reports to the EPA describing the condition of its waterways, the status of its pollution problems, and the estimated cost and environmental implications of reaching full compliance with the act, and recommending further action to restore and maintain the original integrity of indigenous streams, lakes, and rivers. Reports submitted since the act passed into law generally indicate that water quality throughout the nation is slowly improving.

The clearest progress has been made in the reduction of industrial pollution. Among industrial point sources, compliance with Clean Water Act regulations is high, around 90 percent according to some EPA estimates. After first dreading and in many cases actively fighting implementation of Clean Water Act effluent standards, some industries have been pleasantly surprised to find that cleaning up their waste water pollution can be economically beneficial. In numerous cases improved pretreatment of their effluent has led to the added bonus of making it possible for businesses to reclaim and reuse substances essential to their manufacturing processes. Many have also been able to recycle water and use less energy, adding to their savings.

The public sector lags far behind the private sector in cleaning up its effluent. Less than half, perhaps as few as one-third, of the nation's cities and towns are truly in compliance with Clean Water Act regulations.

There are good reasons for this gap. Municipal wastes, especially from large cities, are inherently harder and more expensive to control than industrial wastes in many ways. Whereas the average corporation may only have one factory or one power plant to manage, a large metropolis contains millions of effluent sources, including both domestic as well as industrial dischargers, and must also cope with the added problem of nonpoint pollution caused by runoff into its storm drain system.

Another good reason is that cities in the last decade have been chronically short of capital and the means to raise it. It takes a sizable amount of capital to finance the construction of a major sewage system, and the public has been less and less willing to support the bond measures needed for the local share of these costs. On the federal side there have been regular demoralizing delays in getting the grant funds authorized by the Clean Water Act actually appropriated and distributed to qualified municipalities, and now those funds have been sharply cut. The overall result has been that the process of planning, designing, financing, and building more effective sewage treatment works has been an agonizingly slow one in most metropolitan areas. On top of all that there have been the inevitable squabbles among local vested interest groups about such issues as where treatment plants should be sited and what technologies they should employ.

Though limited on the whole, the progress that has been made in cleaning up our waters is quite tangible in many areas. Fish swim in rivers in the East where they haven't been seen for a hundred years. Though its lower reaches are still heavily polluted, construction of some seventeen new treatment plants along the upper Cuyahoga has insured that it will never catch fire again. The notorious "dead sea," fifty square kilometers of ocean off New York, caused by highly polluted effluent from the hundreds of cities and factories along the coast, is coming back to life as the number of noncompliant dischargers in the area dwindles to a handful.

But, as always, there are new problems to solve. As conventional and some toxic pollutants from industrial and municipal point sources slowly come under control, attention is turning to the much tougher challenges we still face. Nonpoint contamination from urban and agricultural runoff, as well as from construction, logging, and mining sites, will continue to be major sources of water pollution. Protecting our groundwater supplies from the synthetic organic pollutants that defy the natural filtering action of the earth's crust will continue to be a pressing concern. And we are in the midst of discovering that we may need to impose even tougher standards than we once thought necessary in most municipal treatment plants to combat effectively the persistent infiltration of heavy metals and toxic organics. As of December 1979 an EPA study of municipal plants *in compliance with Clean Water Act regulations* found in the eight facilities evaluated that "nine toxic heavy metals (cadmium, chromium, copper, lead, manganese, mercury, nickel, silver, and zinc) are consistently present in all waste waters. . . . Six toxic organic chemicals (benzene, chloroform, methylene chloride, bis (2-ethylhexyl) phthalate, tetrachlorethylene, and toluene) are found in nearly all waste waters including those of domestic origin."[16]

HOME REMEDIES

A note about home water filtration systems, especially those using activated carbon: In a few places, especially where consumers draw their drinking water directly from wells, they may have some value, but in locations served by centralized water utility companies they will not add that much protection to what is already available to consumers from the utility. And they do have limitations. For one thing, only the larger ones offer sufficient absorption. Smaller units that attach to the faucet are inadequate. And the larger units are, of course, expensive.

Second, carbon filters are not effective at removing certain inorganic contaminants, especially minute asbestos fibers.

Third, over a relatively short period of time—a matter of months—carbon filters, even in larger units, lose much of their absorptive capacity. Most organic contaminants of concern, such as chloroform and TCE, will be successfully absorbed to some degree for a while, but after two to three months, depending on the concentration of the contaminant in the water supply, the carbon will become less effective. At that stage a process of selective absorption begins: More highly absorbable contaminants will still be absorbed for a while longer but less absorbable ones will not, and will instead start to pass all the way through the faucet.

Fourth, some studies have found that after a short period of operation, organic substances accumulating in the carbon filter begin to act as a growth medium, actually encouraging the development of potentially harmful bacteria. The consumer may unwittingly be trading a measure of increased protection from organic contaminants for decreased protection from bacterial diseases.

One other subtle drawback to the use of household carbon filters is that carbon is so good at removing bad tastes and odors from water, making it appear even fresher and tastier than regular tap water, that it can mislead a consumer into assuming that the water is safe just because it looks, smells, and tastes safe, when in fact the carbon filter may no longer be screening out invisible, odorless, and tasteless organic contamination very well at all. And the only way a consumer who is relying on carbon filtration as his or her sole protection against organic contamination of water can check to be sure that the filter is still working is to draw a sample of water from the tap and have it analyzed by a competent lab—not an inexpensive procedure.

6

HAZARDS IN THE FOOD SUPPLY

Nowhere is the mixed nature of the blessings that industrialization confers more evident than in our diet. Compared to what it was like two hundred years ago, the food we eat today is more abundantly produced, more successfully distributed and stored, more easily prepared, more varied, and more available to more people than ever before in history. It is probably also more nutritious and may even be more palatable, though that, of course, is a matter of individual taste. There's little doubt that for those of us fortunate enough to be living in the United States—despite the rising costs of food—it is also far less expensive than it is virtually anywhere else in the world. We spend, on the average, less than one-half as much of our incomes for food than the rest of the world does.

On the other hand, for all of the commendable improvements that industrialization has brought to our diet, it has also brought fundamental changes in its nature that may not be so beneficial. Compared to what the average American ate two hundred years ago, we consume more fruits and vegetables, which is good, but we also consume fewer complex carbohydrates, especially in the form of higher-fiber grains and legumes, more sugar, more salt, more saturated fats, more animal protein, more synthetic additives, and more residues of the synthetic chemicals used to fertilize our crops, regulate their growth, protect them from pests, and keep our livestock fat and free of diseases, all of which may not be so healthy.

Along with changes in our diet have come profound changes in who is making the critical decisions about what we eat. Fewer and fewer of us choose to prepare our own food in our own kitchens from raw ingredients that we grow or at least select ourselves. Instead we much more commonly choose to eat what someone else has prepared for us in a food-processing plant hundreds of miles away, or has grown in a field thousands of miles away, perhaps even in a different country. Like so many other aspects of our culture, food has become a mass-produced commodity. This is not necessarily bad—it may even mean that our diet is all the better for it—but it also means that, as individuals, we are exerting less and less control over the contents of what we put into our bodies.

SUGAR

"Sugar" is a generic term referring to more than one hundred different substances, the most common of which is sucrose—table sugar. Once ingested, sugar is broken down by the body's cells, where it is burned as energy, or stored in the liver as glycogen. When the body needs more energy, the liver changes stored glycogen back into glucose and releases it into the bloodstream. To use sugar as a fuel the body must also be supplied with adequate insulin; the disease caused by insufficient or ineffective insulin in the body is known as diabetes mellitus.

Sugar is now the most common substance added to our food.* We are eating more of it than ever before, an average of some fifty-eight kilograms per person per year. The single greatest source of sugar in the American diet is beverages—particularly soft drinks, beer, and wine.

According to the FDA there has been a significant shift in our sugar consumption patterns over the past few decades. Although we are eating more sugar than ever before, less and less of it is sugar that we add to food ourselves. As we cut down, food manufacturers pour it in, and the result is a net increase in our consumption. Today, almost two-thirds of the sugar we ingest—so-called "hidden" sugar—has already been added to our food before it ever gets to us.

Sugar is not inherently toxic. It adds energy (4 calories per gram) to our diet, is tasty, and does help prolong the shelf life of most foods that it is added to.

So what's the harm of this nontoxic energy source in our food? The most common chronic disease in industrialized societies is tooth

*Legally, neither sugar nor salt is considered an additive. Most nutritionists ignore this technicality, however.

decay—only 2 percent of us escape it completely. The $10 billion per year that we spend to have our cavities fixed is in large part caused by our excessive sugar intake, as is the weight problem that one-third of us struggle with. Sugar has also been repeatedly implicated as a contributing factor in various gastrointestinal disorders, diabetes, and hyperactivity in children, but the scientific evidence on these issues is still inconclusive.

Food technologists may have finally succeeded in developing a better substitute for sugar. In July 1981 the FDA announced that an artificial sweetener called aspartame, manufactured by the G.D. Searle Corporation, had finally been approved for public consumption. Aspartame is far sweeter than sugar, gram for gram, and does not have the objectionable aftertaste that some people notice in saccharin-flavored products. It also contains far fewer calories than sugar, and does not cause cavities.

SALT

Salt—or more correctly, sodium chloride—is an essential ingredient in the diet and has a long and illustrious career as a food preservative. Salt was so important to ancient cultures that they literally banked on it: trading in salt determined the major trade routes in Europe, the Mediterranean, and Asia centuries before Christ was born, and salt was commonly used as a form of payment for services rendered (a practice from which our term "salary" is derived). Without reliable staples like salt pork and beef jerky, the discovery of the New World and the steady settlement of the American frontier would have been infinitely more difficult.

However, like anything else, even such essential substances as vitamins, minerals, amino acids, water, and oxygen, salt in high concentrations is toxic. Descriptions of the vomiting and coma caused by "hypernatremia" (acute salt poisoning) abound in medical journals; according to some sources the condition has a mortality rate of over 50 percent. Studies of animals fed a high-salt diet have repeatedly documented its adverse effects, including abnormal swelling of tissues, constriction of blood vessels, and interference with the kidneys' ability to regulate blood flow properly. High levels of salt in the blood decrease its ability to pass through cell walls and make it harder to pump, forcing the heart to work harder.

Even at levels common in the American diet, salt can exacerbate a dangerous predilection for high blood pressure and stroke, especially in blacks and older people. Hypertension is a serious and underestimated health problem in this country, since if not treated it can lead to heart disease, stroke, or kidney disease. Some hypertensive individuals can return to normal by simply reducing their salt intake—all benefit from cutting down on salt in their diet.

Epidemiological studies of preindustrial populations exposed to a high-salt diet consistently note their tendency to become suddenly obese and hypertensive. For example, in 1977 an English physician, Dr. Hugh Trowell, an associate of Nathan Pritikin's, reported to a hearing of the Senate Committee on Nutrition and Human Needs that African tribes accustomed to a diet of 2 to 3.5 grams of salt per day immediately began to experience problems with obesity, hypertension, strokes, and heart disease when inducted into the army and fed a "Western" diet containing approximately 16 grams of salt a day.[1] In 1981 the average American consumed some 6.75 kilograms of salt, or 18.75 grams each and every day. Most medical authorities would agree that 2 to 2.5 grams a day (about a teaspoon) is plenty.

The reason we ingest so much salt is that so much of it is added to the processed foods we eat. Salt is a good, inexpensive preservative—about six times as effective as sugar—and a great seasoning, especially if you're already used to the taste of salty food. However, the amounts used in many processed foods far exceed the requirements of storage or taste. According to USDA figures, one cup of fresh peas or corn contains only about 1 milligram of salt. A cup of *canned* peas or corn can contain as much as 400 milligrams of salt. One brand of processed frozen chicken breast contains five times as much salt as a regular unprocessed frozen chicken breast. For many years baby food manufacturers added liberal amounts of salt (and sugar) to their products because they tasted better to mothers sampling them to see if they were warm enough for baby to eat. (Studies of infants showed that they didn't much care whether their food was salted or not.) Today, some baby food companies have eliminated added salt (and added sugar) in all but a few of their products.

If you're trying to avoid excess salt in your diet, don't eat things like pretzels, smoked meats, ham, bacon, cheese, pickles, olives, peanut butter, salted crackers, canned soups, soy sauce, mustard, ketchup, Worcestershire sauce, and many more such items.

Labels are some help, but not much. Ingredients are listed in order of their relative amounts in the product, with the highest first, the next-highest second, and so on. For example, the label on one brand of soy sauce reads: "Hydrolized protein extracted from corn and soybeans, water, salt, corn syrup, caramel color." Salt is the third most abundant ingredient in this product. It would make much more sense and be much more useful for people who have to follow low-salt diets if the actual amount of salt in milligrams were spelled out on the label, a reform that consumer groups have been urging on the FDA and the food industry for years.

FIBER

"Fiber," also known as "crude fiber" and "dietary fiber," is nothing more than the nondigestible parts of the plants we eat. Cellulose—those long strings on celery—is one common example of fiber. As recently as a century ago our diet was much richer in fiber than it is now, primarily because processing food tends to remove it. At the turn of the century we were eating about twice as much fiber as we are today, primarily in the starchy meat of foods like potatoes.

A good example of the loss of fiber in our diet is white flour. Before the twentieth century, milling removed relatively little of the bran fiber in a kernel of wheat. The resulting flour had more bran in it than what we eat today, but it also spoiled a lot faster, and spoiled grains can be quite hazardous, since a natural, virulent carcinogen known as aflatoxins can form in the mold that grows on them. In the late nineteenth century the advent of special milling and sifting techniques made it possible for the first time to eliminate most of the bran, producing a lighter flour that kept much longer than before. Whiter flour was also more marketable because of its traditional associations with scarcity and status. Of course since this flour wasn't really white, but more of a yellowish color, bakers found they still needed to bleach it, as they do to this day, using acetone or benzoyl peroxide, substances that some nutritional researchers feel have not yet been tested enough to guarantee their safety.

Also, since the new milling techniques caused the loss of some three-fourths of the nutrients along with the bran, bakers learned in later years to make up for the loss, "enriching" this enfeebled bread by adding nutrients like riboflavin, niacin, thiamine, and the old standbys, salt and sugar.

Still later, baking companies found that adding a small amount of calcium propionate, a harmless compound, would help prevent bread from getting moldy too soon. Ironically, this last step canceled some of the need to refine flour so highly in the first place, but by then the American consumer had been hooked on white bread and food industry lobbying had succeeded in passing laws making it illegal to mill flour with higher bran contents.

Because the incidence of a variety of intestinal disorders—colitis, diverticulitis, colon cancer, hemorrrhoids, and so on—has risen in industrialized nations as the level of fiber in their diet has dropped, nutritional researchers speculate that there may be a causal relationship between these two trends. As with salt, this theory has received some support from epidemiological studies of preindustrial cultures, where fiber in the form of beans, bananas, potatoes, and whole grains and rice is still a prominent part of the diet and where many of these same GI diseases are virtually unknown. Researchers speculate that fiber may help protect man from GI disorders by speeding up the process of elimination, thus reducing the time that natural and synthetic toxins in the feces are exposed to the absorbent walls of the intestine.

This speculation about the value of fiber has also received some support from recent animal studies showing that a wheat bran diet seems to suppress the growth of deliberately induced colon tumors,[2] but the results of such studies are not consistent.

Finally, some researchers theorize that fiber may play a role in preventing heart disease. One view, based on epidemiological evidence that cultures with a high-fiber diet tend to have low blood cholesterol levels, holds that fiber washes a cholesterol byproduct known as bile acids out of the body, triggering the conversion of more cholesterol to replenish lost acids, thus lowering blood cholesterol levels.

So far the suggested connection between lack of fiber in the diet and industrialized disease patterns is still speculation. As is true of virtually every epidemiological study, it is hard to say exactly what is responsible for the disparity in GI disease levels between industrial and less-developed cultures. Differences in consumption of meats, fats, and alcohol, as well as broader life-style differences, may also be at work. Critics of the profiber position point out that there is no documented requirement for any fiber at all in the human diet, as there is for other animals, namely rabbits, guinea pigs, and cows. Some attribute the popularity of fiber to the misguided conviction that a daily bowel movement is essential for good health, a belief that has no basis in medical fact. Some even note that high levels of fiber in the diet may be harmful. The sheer bulk of a mass of indigestible fiber can cause painful and even dangerous intestinal swellings. Also, since fiber absorbs water like a sponge, it can leach vital minerals out of the body, a particular hazard for growing youngsters.

Based on guinea pig studies, one fiber researcher, Dr. George Briggs of the University of California at Berkeley, suspects that fiber's true value may have nothing to do with the elimination of harmful substances from the body. Briggs suggests just the opposite, that fiber may contain a powerful but hitherto unidentified vitamin-like substance, one whose chemical nature and health effects remain to be fully analyzed. In the meantime, Briggs points out, until the fiber issue is resolved, most of us can get all the fiber we could conceivably need by just regularly including a moderate amount of fresh fruits and vegetables in our diet.

CHOLESTEROL

Cholesterol is a natural substance produced in the body by the liver and present in many foods, especially animal fats and dairy products. Over the last few decades it has been repeatedly implicated as a risk factor in the development of cardiovascular disease, but its role is still a highly ambiguous one. Most medical authorities believe that high cholesterol levels in the blood contribute to the buildup of fatty "plaque" along the walls of the blood vessel, a process known as atherosclerosis. Atherosclerosis can eventually clog arteries so badly that blood no longer circulates through them. If the clogged arteries happen to be those that feed blood to the heart, then loss of circulation causes loss of the oxygen needed to keep heart tissues alive, and the result is pain, diminished heart function, and, in severe cases, heart attack. If the arteries are those that feed blood to the brain, the result is stroke.

Cholesterol is thought to be one of the substances most heavily involved in the buildup of plaque, but this is a controversial topic. It is clear that eating more fruit, vegetables, lean meat, fish, poultry, and nonfat milk in place of hamburgers, hot dogs, bacon, sausage, cheese, eggs, french fries, potato chips, and ice cream does reduce cholesterol levels in the blood for most people, but whether reduced cholesterol levels reduce the risk of atherosclerosis is still an unresolved issue.

For one thing, the human liver produces some two-thirds of the cholesterol found in our blood, regardless of what we eat. For another, cholesterol is accompanied by three different kinds of fatty proteins—high density (HD), low density (LD), and very low density (VLD)—and the HD type appears to *lower* rather than raise blood cholesterol levels. So we may soon find that the advice we're hearing from doctors is no longer a simple matter of "avoid animal fats and dairy products" but a more complicated "skip the pastrami but eggs are okay."

FOOD INFLATION

One of the most unhealthy trends in the contemporary American diet has little to do with toxic contamination. In many ways, it is much more reminiscent of our current economic troubles than any medical problems.

Sugar, white flour, and fats, especially vegetable oils, have combined in an unholy trinity to rob us of valuable nutrients. All of these highly refined substances provide calories and little else—few vitamins, minerals, or other essential nutrients. Yet the average American's diet relies more and more heavily on them. As a result some nutritionists calculate that as much as 50 percent of our diet has become "empty" calories, and that that 50 percent is providing us with only 10 percent of the nutrients we must have to stay healthy. As more and more of our diet is made up of the processed foods that these three substances figure so prominently in, other, healthier foods are crowded out, and as a result the nutritional value of our diet declines. For people who regularly drink much alcohol—another source of empty calories—these figures may be even worse.

We seem to be suffering from a strange kind of food inflation, one in which the calorie level of our diet keeps rising at the same time as its nutritional value keeps falling. What makes this trend especially unhealthy is its contribution to obesity. In an age when machines have taken over practically all of the actions in which we once expended calories, and most of us lead very sedentary lives, what we need is fewer calories in our diet, not more. The more calories we eat that we don't burn up, the fatter we get, and the greater grows our risk of developing heart disease, diabetes, and a number of other degenerative diseases.

FOOD ADDITIVES

Food additives have their origin in the need to store food for long periods of time, whether to transport it over long distances or to maintain a regular food supply between harvests. This is an ancient need, one that mankind first met by such simple practices as salting, smoking, drying, and otherwise removing from foods the moisture that provides a growth medium for bacteria and molds. Without such protective measures food can quickly spoil and become moldy, rancid, unappetizing, and even quite toxic. Botulin staphlococcus, salmonella, aflatoxins, and a host of other natural food pollutants are among the nastiest poisons known to mankind.

But the process of treating foods so as to make them last longer and protect them from these contaminants imposes its own health risks. Smoking a foodstuff, we now realize, can introduce a number of potentially harmful carcinogenic and mutagenic compounds into it, such as benzoapyrene and formaldehyde. And, as we have seen above, salted foods are often associated with hypertension and other cardiovascular disorders.

In preindustrial societies the amount and variety of foods that can be preserved by salting, smoking, and drying is limited. One of the main reasons villages in such societies tend to remain villages rather than growing into towns and cities is that their size is limited to the number of people who can be fed by the yield from the immediately surrounding countryside and the small stores of preserved foods they can accumulate.

Industrial progress brings about a cycle of growth in such societies by introducing both the means to feed many more people much more efficiently and the need to concentrate them in fewer places more densely. For example, along with one typical industrial innovation, refrigeration, comes a tremendous expansion in the food storage capability of a culture, as well as the need for many more people to work in refrigerator factories and food-processing plants, to transport raw materials to these factories and plants, to service the refrigerators, to run grocery stores, and to maintain the power and transportation networks that make the spreading use of these innovations possible in the first place.

Such industrialization of food production and distribution would not be possible without thousands of chemical additives, needed to make sure that processed food does not spoil, remains nutritious and appealing to the customer, and is amenable to the techniques of industrial processing. In short, industrialization of food production brings about an intense, pervasive infusion of chemical additives into the food supply.

Today, each American consumes two to three kilograms of about three thousand different additives a year. Some of them are clearly beneficial. Iodine, for example, added to salt has virtually eradicated a serious thyroid disorder known as goiter in the United States, which as recently as 1924 could be found in as many as one-third of the schoolchildren in some parts of the country.[3] Vitamin A added to margarine and some other foods helps protect us against the eye diseases that can arise from its absence in the diet, and vitamin D added to milk helps prevent rickets, a serious bone disorder.

Other additives are clearly useful, but not necessarily healthful. Without calcium silicate in our salt to absorb moisture it would cake up in the shaker. Without emulsifiers in mayonnaise and salad dressings the oil would separate and float to the top of the jar. Without a spraying of maleic hydrazide too many potatoes would sprout before we could eat them. A dose of ethylene gas, given off naturally by many fruits as they mature, ripens honeydew melons in time to be marketed promptly.

Other additives, especially most artificial colors, are useful only in a cosmetic sense. They make our food look the way the $2-billion-a-year food advertising industry wants us to believe it should.

Since the turn of the century dozens of once-common food additives—including substances like charcoal, methyl alcohol, and formaldehyde—have been found to be unsafe and have been banned from our food. Others still in use today are considered questionable by many scientists and some are in the process of being reviewed by the federal agency responsible for assuring the safety and purity of our food, the Food and Drug Administration.* We discuss three of these additives—nitrite, saccharin, and estrogen—below.

NITRITES

Nitrites are an inorganic salt derived from nitrogen, specifically from nitrous acid. Nitrites are a valuable additive in cured meat, especially pork products like ham, bacon, and sausage, because they help prevent the formation of botulin, one of the deadliest food poisons known to man. Botulism is so poisonous that a thimbleful mixed with water theoretically has the potential to kill everyone in the world. Coincidentally, nitrites help make meat a nice, attractive, reddish-pink color, a by no means negligible bonus in the packer's eyes.

In foods nitrites pose a problem because they can combine with stomach acids to form a very potent class of carcinogens known as nitrosamines.

In 1978 the FDA learned the alarming news from a scientist it had contracted with to study the carcinogenic potential of nitrite that

*A detailed discussion of even the more significant additives under suspicion would far exceed the scope of this chapter. The interested reader may want to consult the comprehensive treatment of this whole topic in Michael F. Jacobson's *Eater's Digest: The Consumer's Factbook of Food Additives* (Anchor, New York, 1972).

it appeared to be a clear cause of cancer of the lymph system in rats. The evidence looked so strong at first that the agency immediately began to take steps to order the substance barred from foods. On August 11, 1978, the FDA announced in conjunction with the USDA that it was planning to phase out the use of nitrites in food. The implications were staggering: A total of some 4 billion kilograms of meats, poultry, cheese, and fish (one-tenth of everything we eat) would be affected, and without an adequate substitute for nitrites the risk of botulism faced by nearly everyone would suddenly become acute.

However, within months of this announcement the agency had learned some encouraging but, under the circumstances, highly embarrassing news from outside reviewers of the study on which it had based its decision to move against nitrites. Although it still seemed clear that nitrite caused some kind of malignant tumor growth in rats, it was no longer clear just what it was or how significant it was. Moreover, other evidence had surfaced that seemed to indicate that other common substances in foods, such as the ascorbic acid (vitamin C) found in, say, a glass of orange juice, effectively blocked the formation of nitrosamines in the human stomach. A little more than two years after the phase-out was first announced, the FDA rescinded it.

Today, the nitrite issue is still unresolved. Although it now seems less likely to FDA scientists that nitrites are a direct, significant cause of cancer, there's still no doubt at all that they do, under certain conditions, combine with amines to form nitrosamines. There's also no doubt that nitrosamines are a virulent carcinogen in animals, but how common their formation is in foods or the human body because of the use of nitrite additives is not at all clear. However, some meat companies, seeing the handwriting on the wall, have begun to market cured meat products containing no nitrites at all, and there has been no discernible increase in botulism as a result.

ADDITIVE REGULATION

The FDA, an agency of the Department of Health and Human Services, is responsible for regulating cosmetics, drugs, medical devices, and foods, including food additives. Its authorization to do so comes from a host of laws, including what is probably the oldest environmental health legislation in the country, the Food and Drug Act of 1906. Not until passage of the Food, Drug and Cosmetic Act of 1938, however, was the FDA empowered to take hazardous foods off the market, if necessary, and not until the late 1950s was it granted the authority to regulate food additives and coloring agents.

Amendments to the act in 1958 and 1960 mandated FDA regulation of additives and colors, and also introduced an extremely important shift in the presumption of food safety. Before then it was up to the FDA to prove that a product was potentially dangerous. After passage of these amendments new food products and additives were presumed to be guilty until proven innocent, and it was up to the manufacturer who wanted to introduce a new product to prove that its consumption would not harm the public.

The testing process, which usually takes years to complete and is quite expensive, involves feeding at least two different animal species varying doses, including extremely large ones, of the additive over the life span of the animal species being studied. (A general description of this process appears in Chapter 1.) The manufacturers then assemble the data from these tests in a food additive petition and submit it to the FDA. If the test results look acceptable to FDA scientists, the agency establishes guidelines for the additive's use in foods and announces its intention to approve the substance in the *Federal Register.** After sixty days, during which time anyone can object to the proposed use of the additive, the manufacturer can then begin to use it.

A major problem with this and other similar regulatory approval processes is that only the manufacturer and regulator have access to test data. Due to industry fears of revealing trade secrets, laws requiring health and safety testing invariably also contain confidentiality clauses that allow manufacturers to indicate which portions of their petition they do not want publicly disclosed. Unfortunately, though these clauses are intended to keep information on yields, costs, and manufacturing techniques secret, they tend to be interpreted so broadly by both the manufacturer and the regulator that they keep health and safety data secret, too, denying outside scientific and public interest reviewers the opportunity to evaluate their public health implications independently.

The 1958 and 1960 amendments also introduced the concept of a "Generally Recognized as Safe" (GRAS) list. Because hundreds of the additives and colors that came under FDA regulation for the first time in those years had been used for decades without apparent

*The *Federal Register,* the official news publication of the executive branch of the federal government, contains all new rules or changes in old rules proposed by regulatory agencies, among other things.

harm to the public's health, it was felt that suddenly requiring the food industry to conduct exhaustive testing on all of them would place an unfair, time-consuming, and expensive burden on it, one that would also remove many clearly useful additives from public use for some years while the tests were being conducted. So a list of some seven hundred GRAS substances was compiled and certified for temporary use until the FDA could get around to checking the available scientific information to find out if any of them seemed questionable. This review, which was conducted by a Select Committee on GRAS Substances of the Federation of American Societies for Experimental Biology, did not begin until 1971 and was not completed until ten years later. Of the 450 substances evaluated, 305 were certified as safe. Some were found to be unsafe, and in a few cases, including caffeine, salt, BHA, and BHT, the committee recommended restrictions on their use or further study of their health effects.

The 1958 and 1960 amendments also contained what soon became a very controversial clause on carcinogens in food products. Known as the Delaney Clause, after New York Representative James J. Delaney, who sponsored it, the clause basically states that if a substance is found to be a carcinogen in humans or animals, it shall be deemed unsafe. Strictly speaking, the clause has made it illegal for anyone to add *any* substance to *any* food if there are *any* indications whatsoever that it is carcinogenic at *any* dosage level in *any* animal. This may seem like a good thing, but many scientists and regulators argue that it is impractical and that it has been inconsistently enforced—on more than one occasion both the FDA and Congress have ignored or circumvented it, usually because the substances in question were very weak carcinogens, the evidence of their carcinogenicity in humans was highly speculative, or because the American public made it clear that they preferred to continue accepting the risk of exposure to these substances rather than do without them. The most notable example of this last reason is saccharin.

SACCHARIN

Saccharin, which has no calories and is hundreds of times sweeter than sugar, has been used commercially since 1900 in this country as a sweetener for diabetics, who must restrict their intake of conventional glucose sugars. As the American public steadily became more weight conscious over the last few decades, the use of saccharin and other artificial sweeteners soared, especially in soft drinks. But in 1969 one class of these substances, cyclamates, was withdrawn from the market after it was found that they could cause bladder cancer in test animals, and almost a decade later, after a series of studies indicated that saccharin also causes bladder cancer in mice and rats, the FDA began in April 1977 to move to have saccharin's use in foods and beverages prohibited under the Delaney Clause.

There was an immediate, vociferous public protest. Some scientists agreed that saccharin was hazardous, but others pointed out that the best of the saccharin studies had only been able to show that saccharin was a very weak carcinogen, that only male rat fetuses had shown any significant susceptibility to its carcinogenic properties, and that only enormous doses, five hundred to two thousand times the amount that would be consumed by the typical human consumer, had caused tumors. Furthermore, this latter group noted that even though saccharin's use had been increasing in the United States ever since 1900, deaths from bladder cancer had been steadily falling since then.

Finally, in late 1977, Congress took the matter out of the FDA's hands by passing a law (the Saccharin Study and Labelling Act) ordering the FDA more or less to halt in its regulatory tracks while the National Academy of Sciences reviewed the saccharin data and made recommendations. After studying the tests the NAS agreed that saccharin did indeed appear to be a weak carcinogen in rats, but recommended, instead of banning it, that the Delaney Clause be changed. The reasoning behind the NAS position, with which the FDA itself and most scientists concur, is that the danger of a carcinogen in food should be evaluated by its statistical risk, not by its simple presence or absence in food. We are now able to measure much smaller quantities of many substances than we were in the late 1950s, and in many cases the levels we can detect are so low that it would be senseless to try to banish the compounds we find using these methods just because we are *able* to find them. It makes sense, according to this argument, to calculate the risk of forming additional tumors that carcinogenic food additives can cause, based on knowledge gained from animal studies, and then make sure that risk is kept below one in a million for humans by keeping the concentrations of these additives below the level associated with those odds.*

*The risk is actually even lower than that because these calculations are based on results from the *most* sensitive animals found to be susceptible to the carcinogen, and because they are based on the unrealistic assumption that everyone in an exposed

In 1979 Congress extended the saccharin ban for another two years.

When it expired in 1981, Congress once again extended the moratorium for another two years, despite an announcement from the FDA that it did not intend to move against saccharin. In that same year the House, the Senate, and the FDA all introduced legislation proposing fundamental reforms in food safety regulation, primarily by substituting a risk-assessment process for the Delaney Clause.

ESTROGENS

A certain class of additives—most notably a number of different kinds of antibiotics and artificial estrogens—are added to meats and poultry while it is still on the hoof (or claw). All of these compounds leave tiny but detectable residues in the food we eat. Some also leave significant residues in the ambient environment, deposited there via wastes excreted by animals treated with these substances.

Of all these compounds, estrogens are probably of greatest potential concern since, although they are produced naturally in animals and even in some plants, they have the potential to cause powerful disruption of human reproductive cycles, producing effects ranging from enlarged breasts in men to abnormal menstrual cycles in women. DES (diethylstilbestrol), the most commonly used synthetic agricultural estrogen, also has a checkered history as a pharmaceutical antidote given to women about to miscarry. Used heavily for this purpose in the 1950s and early 1960s, it turned out later to have had an unsuspected side effect: we now know that in some cases DES in these high doses crossed the placental barrier and caused serious teratogenic effects, including decreased fertility and structural genitourinary defects in sons born to DES-treated women, and changes in vaginal mucosa, uterine alterations, and even vaginal cancer in a small number of daughters born to these women.

In 1977 we used nearly thirty thousand kilograms of DES to help fatten the cattle and sheep headed for our tables. Well aware of the potential of DES and other agricultural estrogens to cause serious health effects at high doses, the FDA has aggressively monitored their residues in foods, carefully controlling their use ever since the mid-1970s. Today, there seems to be little likelihood that the

---

population consumes the maximum acceptable dose of the additive each and every day.

minute amounts of estrogens that virtually all of us (even vegetarians) consume in foods will cause us harm.

However, even though its traces in meat and poultry may be carefully monitored, there is another potential source of exposure to estrogens that is not well analyzed at all. DES and its chemical cousins were all in use long before stringent testing for ambient environmental effects was a common requirement for food additives. Consequently, there is to this day no clear understanding of the environmental state of the tens of thousands of kilograms of synthetic estrogens and other substances that pass through our livestock and return to the biosphere via their wastes. Studies are just beginning to be conducted on this topic, and some of their preliminary conclusions are disquieting. Under certain conditions, for example, it appears that DES residues can survive for months in soil and in water supplies, where they are readily taken up by other organisms, including food crops and fish.

PESTICIDES AND THEIR RESIDUES IN FOODS

Pesticides have been known and used for hundreds, perhaps thousands, of years. Until the middle of this century they were mainly derived from inorganic substances, like lead and arsenic, or isolated from plants, like pyrethrum, nicotine, and rotenone. Then, a series of revolutionary discoveries made by European chemists before and during World War II introduced dramatic changes into the pesticide world, ushering in what has become the age of the synthetic pesticide and elevating the pesticide business, now dominated by the world's oil and chemical companies, to the status of a major international political and economic force in the world of the 1980s.

The pesticide production business in the United States is now a $5-billion-a-year industry, one that markets some thirty-five thousand different products, all formulated out of some one thousand basic chemical components registered with the EPA. Although most of its output is still used domestically, the portion sent overseas has risen to 30–40 percent in recent years and by the mid-1980s will in all likelihood be greater than that sold at home. Most of the pesticides used in other countries, especially in the developing nations of Latin America and Southeast Asia, are made in the United States, Europe, and Japan, or are made out of basic chemical ingredients manufactured in these places and then shipped overseas for final mixing into the usable product.

Viewed historically, pesticides seem like an inevitable development in the long evolution of agriculture. Before man began to cultivate food crops, he was a hunter and food-gatherer, simply collecting whatever he could find growing wild that was edible and nutritious. In natural, uncultivated ecologies there tends to be a harmonious balance between plants and their predators. The highly varied vegetation that occurs in these settings helps control pests because it offers only limited food supplies in any one location, usually includes plants that attract pest predators, and forces all organisms—plants, animals, and insects—to compete more or less equally, a condition that favors the evolution of only the hardier, more resistant species and strains.

However, as man learns not only to gather but also to deliberately cultivate food crops, this natural equilibrium is upset. He clears away all growth in a certain area, then plants only a few crops, which he carefully tends, provides with water and nutrients, and clears of weeds. As a result of this shift from polyculture to monoculture, from variety to uniformity, those insect and other pests that happen to prey on the same crops that man selects for cultivation find themselves presented with unnaturally abundant food. And the natural result of a sudden expansion in their food supply is an increase in their population. At the same time, man's consistent eradication of the plants and other conditions that support their natural enemies exacerbates the tendency for their numbers to soar beyond control.

In the meantime man the farmer discovers that the dependable food supplied by monoculture encourages his own population to expand, that, once started, its expansion is not easily stopped, or even slowed, and that its continued expansion requires increasingly larger harvests, which can be produced in only one of two ways: Either by bringing more land under the plow or by increasing the yield from land already under cultivation.

Ultimately the effect is a three-way race between the growth of the pest population, mankind's population growth, and mankind's ingenuity at devising new ways to control pests, manage his population growth, and protect and nourish his crops. We are still very much in the middle of that race today. In some respects we have never yet come close to winning it. In North America alone some three thousand insect pests make a feast on our food every year; entomologists estimate that we still—despite our extensive use of pesticides—unintentionally and unwillingly produce a greater volume of insects each year than we do of any domestic livestock we intentionally raise.

Industrialization intensifies this cycle by making monoculture more efficient and more productive. More land can be cleared and cultivated more efficiently by machines and chemicals than by hand, which means that, for a while, more food is available for more people. But this in turn means, a few generations later, that even more people must be fed by a finite amount of land. The pressure to increase yields from available land by developing even better fertilizers, pesticides, and the plant strains that respond to them best becomes inexorable.

The farmer is on the front lines of this struggle, completely vulnerable not only to pests, but to capricious shifts in both the weather and the market for his crops. Even if today's "farmer" is actually an agribusiness conglomerate, its continued survival is just as easily jeopardized by an early rain, a late frost, or a sudden infestation of beetles. Under such conditions of extreme insecurity it is hardly surprising that when he (or it) finds a chemical that helps him reduce that level of uncertainty by getting rid of a serious pest at a reasonable cost, he will take it up avidly, use it heavily, and become extremely reluctant to give it up for the sake of newer, unproven weapons in his war against pests, no matter how much safer or healthier they may sound.

The term "pesticide" is a broad one, encompassing substances that kill insects, weeds, fungi, algae, worms, mites, and rodents, among other nuisances—anything that threatens man's food supply or other agricultural and industrial commodities, such as lumber. Pesticides are ubiquitous. No one escapes contact with them. The most direct and most hazardous exposures occur among those who manufacture and formulate them, or apply them, or live or work near areas where they are used heavily. But all of us come into some contact with them, whether they are in the form of ambient emissions in the air and water, or of substances, with which we rid our homes and gardens of pests, or of residues in our foods.

CHLORINATED HYDROCARBONS

The single most important factor in the pesticide boom of the last forty years was the discovery in 1939 by a Swiss chemist, Paul Müller, that dichlorodiphenyltrichloroethane, otherwise known as DDT, was a superb insecticide. Working at the laboratories of a major European chemical firm, Geigy Ltd., Müller and his colleagues were trying to create a better moth repellent. DDT, which had first been synthesized some fifty-five years earlier, was only an obscure chemical curiosity until Müller routinely tested it on flies and found that it was an almost perfect insecticide: lethal to insects, long lasting, and relatively nontoxic to mammals.

Geigy patented DDT in 1940. By 1943 both the Allies and the Germans had, with the cooperation of the Swiss, obtained the formula. (To preserve its neutrality Switzerland made certain that each side received copies of it.) Soon both powers were feverishly producing DDT for military use in the tropics to protect troops from such devastating insect-borne diseases as malaria, typhus, and dysentery. In 1943 DDT was successfully used for the first time on a mass scale by American forces in Naples, Italy, to prevent a nascent typhus epidemic. Delousing powders made with DDT soon became a fact of life in refugee camps throughout war-torn Europe. By 1944 factories in the United States were producing some 1.3 million kilograms a year of DDT. After the war it became publicly available for the first time, production soared, and DDT was hailed around the world as a chemical miracle. In 1948 Müller won the Nobel prize for unlocking its secret.

For the next two decades DDT and its numerous chlorinated hydrocarbon insecticide (CHI) cousins were used with great abandon on nearly everything, everywhere. CHIs were used to control boll weevils in cotton, cutworms in rice, crickets in grain, cabbage moths in cabbage, and thrips in potatoes, to name only a few of their applications. They were also used—and still are in many parts of the world—to help control malaria, yellow fever, sleeping sickness, "river blindness," and a host of other debilitating diseases. In India alone, DDT use caused the incidence of malaria to drop from 100 million cases in 1935 to 150,000 cases in 1966.

In the mid-1950s, however, ornithologists and amateur birdwatchers began to notice that strange things were happening to birds living in areas where DDT and other CHIs had been used. As a federal report later put it, they would typically "fly poorly or flutter along the ground, then become totally disabled, undergo convulsions and die in a very stiff position, with legs extended. Such deaths have been observed at many locations and the evidence links losses to DDT, dieldrin, and other insecticides."[4] It also soon became clear that certain bird species, especially bald eagles, peregrine falcons, ospreys, and brown pelicans were suffering drastic population declines. Further investigation revealed that accumulated DDT in their bodies was interfering with the production of the calcium compounds that make their eggshells hard. As a result, afflicted birds were laying eggs with no shell at all or with shells so thin they were crushed the first time a nesting bird tried to sit on them.

True to the dictates of the Sideslip Factor, it was the very persistence of DDT in the environment that caused its undoing. In 1962 biologist and author Rachel Carson wrote the book, *Silent Spring,* that brought the innocent enthusiasm of DDT's boom years to a quick end. As Miss Carson was among the first to point out, because CHIs bioaccumulate in the fat tissues of living organisms, they move up the food chain, becoming more and more concentrated in the tissues of higher-order predators, including humans. In 1957 dead western grebes were found with DDT levels as high as 1,600 ppm in their tissues, because of this process. Samples of the fish they customarily

Figure 19

fed on were found with DDT levels in their tissues as high as 2,500 ppm.[5]

Silent Spring's impact on not only the general public but also the governmental and scientific attitudes of the early 1960s was so profound that one pesticide expert we talked to distinguished between "B.C." and "A.C." eras—Before Carson and After Carson. Seven years after the publication of *Silent Spring* the Department of Health, Education and Welfare recommended that CHI pesticides be phased out by 1971. Over the entire decade of the 1970s, beginning with DDT in 1972, most of them gradually were.

Today, relatively few CHI compounds are still in use, and those that are employed are subject to very stringent application conditions. Aldrin/dieldrin, heptachlor, and chlordane are still used in termite control, and aldrin/dieldrin is also used for mothproofing of clothes. Chlorobenzilate is used on citrus crops in Florida, Texas, California, and Arizona, under rigorous safety precautions. Toxaphene, yet another CHI, can still be legally used on cabbage and lettuce, but only before the heads start to form.

Endrin, suspected of causing reproductive disorders in women, is still permitted for some uses on cotton, grain, apples, and sugar cane, However, the EPA has forbidden pregnant women to work with or near endrin, and even those who are allowed to work around it must wear certain kinds of protective clothing, namely "long-sleeved shirts and long pants of a closely woven fabric, and wide-brimmed hats." Critics of this regulation argue that it constitutes illegal discrimination against women. Just how enforceable such specialized conditions for CHI use are is questionable. Critics claim that CHI use is never adequately monitored, and point out that the EPA does not now and is never likely to have sufficient inspection and enforcement staff to insure that these use conditions are actually complied with. These people feel that all CHIs should be banned, despite their small but significant differences in effect.

The only reason any CHIs at all are still in use is that the EPA has judged that in some cases, with certain crops and pests, there is no comparably effective substitute and that withdrawing the CHI from the market would cause a significant economic hardship to the farmers involved and to consumers of their products. Even so, the EPA is quick to declare that if new evidence appears to show that any of them pose more of a hazard than previously believed, as may now be happening with toxaphene, even greater restrictions—up to and including cancellation of their registration—will be implemented.

Though DDT itself is no longer used here, some 15 to 20 million kilograms of it per year are still made by a Los Angeles firm for shipment overseas.

Caused primarily by concern about the decimation of bird populations, the 1972 DDT ban also stemmed from the discovery in 1970 that DDT and most other CHIs were carcinogenic in lab animals, a particularly disturbing finding in light of their extreme persistence in the environment and in the food chain. Unlike most later pesticides, which biodegrade quickly after application, DDT and other CHI residues are so stable they tend to remain in the environment for decades, contaminating our air, water, and our food, especially meats and dairy products. Furthermore, they are largely unaffected by such conventional purification practices as water chlorination.

At the height of its use in the mid-1960s, we were producing well over forty-five thousand metric tons a year of DDT, one-fifth of which was used in the United States. We were ingesting about 200 μg of DDT each day in our food.[6] However, despite heavy, chronic exposure to DDT and other CHI pesticides, there is no evidence to date clearly linking it to the appearance of large-scale public health disorders of any kind. Furthermore, since its ban in 1972, DDT levels in the tissues of inhabitants of the United States have been steadily dropping. Between 1970 and 1976, DDT levels in human tissue samples decreased from an average 7 ppm in whites and 12.5 ppm in blacks to 4.5 ppm and 7 ppm, respectively.[7] Though persistent DDT residues are still causing hormonal aberrations among some species, the populations of pelican, falcon, osprey, and eagles that were once threatened with extinction by DDT are on the rise again.

Farmers liked DDT not only because it was effective, long lasting, and would kill practically any insect exposed to it, but also because it was cheap. A kilogram of DDT cost something like forty cents, far less than the ten to twelve dollars that most other insecticides used today cost.

But even if DDT had not caused biomagnification problems we still wouldn't be using it much today for the simple reason that insects rapidly develop resistance to it, as they do to most insecticides. In 1981 it takes something like seventy thousand times more DDT than it took thirty years earlier to kill a tsetse fly because of the development of resistant strains.

At first farmers met the problem of resistance by increasing the amounts of a pesticide that they applied to their fields, but this quickly led to what has been called the "pesticide treadmill"—heavier and heavier applications trigger increasing insect resistance, which in turn demands even heavier applications, and so on. Today it has become clearer that a more effective response to resistance in insects is to rotate pesticides, much like rotating crops, so that insects don't have a steady chemical stress to develop resistance to, but even this tactic has its limitations.

ORGANOPHOSPHATES

The banning of DDT and its chemical cousins opened up a need for environmentally safer substitutes. To a great extent this need was met by a group of chemicals known as organophosphates, or OPs.

Phosphorus is one of the basic components of life, found in the protoplasm of all living cells. Research into synthetic organic phosphorus compounds appeared as early as 1820 in France and was well advanced in some respects by the 1930s. However, the major breakthrough that brought OPs to the forefront of postwar pesticide trends occurred in 1944 in a German laboratory in connection with the Nazi search for more effective nerve gases. A German scientist, Gerhardt Schrader, was able to synthesize successfully, among a number of other deadly substances, the compound we now call parathion, a prototype of many current OP pesticides. After the war Nazi OP research appropriated by the Allies formed the technological basis for the development of what has now become one of the major areas of pesticide research and development. Well over one hundred different OP compounds have been synthesized so far, and, theoretically at least, the total possible OP permutations waiting to be developed runs into the millions. In the mid-1970s the United States was using over 50 million kilograms of OP pesticides per year.

OPs are less harmful to the environment than chlorinated hydrocarbons because most of them are far more biodegradable. Exposure to sunlight, water, heat, and oxygen, among other things, breaks them down into harmless chemical byproducts within days, or in some cases, weeks. Furthermore, most OPs are water-soluble, so even if small amounts somehow do find their way into the food chain, they usually do not linger in fat tissues, as DDT does, but are excreted in the urine. There are a few OPs that rival CHIs in persistence, but for this reason they are used only under very restrictive conditions, if at all.

Despite their high biodegradability, many OPs are exceptionally toxic immediately after application, and in some cases for weeks af-

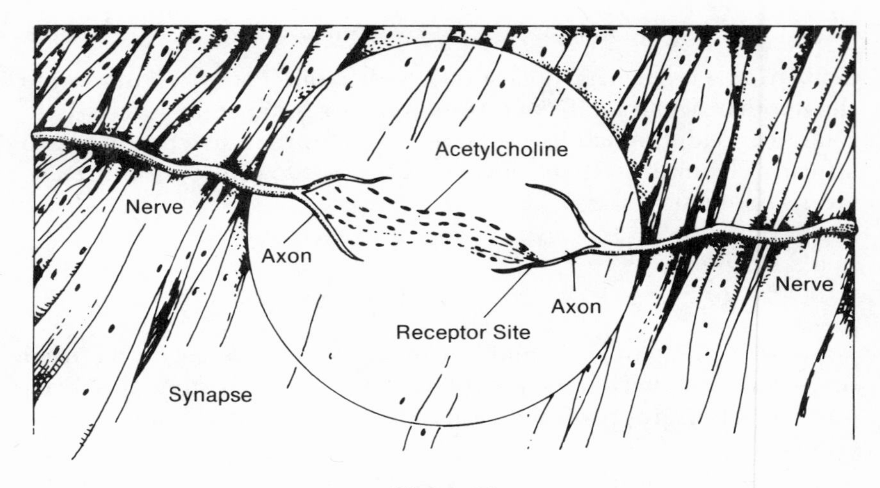

Figure 20

terward. On an acute basis, parathion, the most widely used OP in this country, is many times more toxic than DDT and has probably caused more human deaths than all the CHI compounds together. In some countries parathion is considered too dangerous to applicators to be used at all.

The toxicity of OP compounds is caused by their effect on the nervous system. The nervous system is full of microscopic gaps, called synapses, that function something like circuit breakers. In order for an impulse to cross a synapse from axon to axon, a natural chemical called acetylcholine must be present. And conversely, in order to keep an impulse from crossing a synapse, it must be absent. Muscle actions like raising a finger require first that an impulse be sent along the nervous system, commanding the finger to move, and second that the impulse be stopped. If it isn't stopped, and the muscle continues to receive commands to act, it soon becomes overstressed and eventually quits functioning. If it happens to be a muscle that helps operate a vital organ system, the consequences of this lapse can be fatal.

The natural substance that helps stop transmission along the nervous system by wiping out acetylcholine is known as acetylcholinesterase (ACHE). OPs permanently destroy ACHE in the body. The victim of OP poisoning goes into convulsions, and if the poisoning is severe enough, dies, most often of respiratory failure.

OPs can be inhaled, ingested, or absorbed through the skin. Because OPs have no telltale irritating effects, their absorption into the body is often not noticed by the victim. The signs of OP poisoning range from a mild flulike condition to death. Milder stages are characterized by headache, blurred vision, weakness, nausea and vomiting, heavy sweating and salivation, and abdominal cramps. More advanced stages are marked by muscle twitches, tremors, loss of coordination, especially when walking, constricted pupils, dimming vision, increased urination, and a feeling of tightness in the chest often mistaken for heart attack. Extreme OP poisoning is characterized by stupor, loss of consciousness, coma, low blood pressure, shallow respiration, irregular heartbeat, seizures, and death, usually caused by respiratory failure in human victims.

Much of what we know about OP toxicity in humans is based on acute poisoning incidents involving farm workers. The major route of exposure in these cases is usually inhalation and skin absorption, rather than ingestion. For example, on July 10, 1980, at two A.M., a cauliflower field in Salinas, California, was sprayed with two OPs, mevinphos and phosphamidon. Both are highly toxic, with an LD_{50} of less than 50 mg/kg. They break down rapidly after application, but the EPA requires a waiting period of forty-eight hours before workers can enter fields that have been sprayed with them.

For some reason on this occasion the reentry time was overlooked. Twenty-two workers began entering the field the next morning at six thirty. Within minutes they began to develop headaches and blurred vision. Within hours all were feeling dizzy, nauseous, and weak. When two of them finally collapsed, most headed for local hospitals.

Of the twenty-two, nineteen are known to have received medical treatment, and they all revived quickly. Weekly medical exams, however, indicated that some of their symptoms, particularly vision problems, lasted for as long as three months after the incident.

One of the most common and least toxic OPs is malathion. Patented in 1950 by the American Cyanamid Company, malathion is a general purpose compound used on a wide range of fruits and vegetables to manage a broad spectrum of insect pests. It has been used extensively by the World Health Organization to help control malaria-carrying mosquitoes.

Malathion is valuable because it is a highly selective pesticide: though extremely poisonous to insects, its toxicity to mammals is very low. Experiments with rats, for example, demonstrate that ingestion of as much as 1,000 ppm malathion for as long as nearly two years causes *no abnormalities whatsoever.* The acute oral LD_{50} for malathion in rats is 1,375 mg/kg; by comparison, parathion's acute oral LD_{50} in rats is 2 to 13 mg/kg. One physician we know uses this graphic example to teach student nurses the difference in the relative toxicity of the two substances: "I have two squirt guns. One is filled with malathion, the other with parathion. If I hit you with five squirts of the parathion, you'll fall over dead. If I hit you with five squirts of the malathion, you'll walk away and feel no ill effects at all." Malathion has also been extensively tested for carcinogenic effects—to date there is no evidence that it causes cancer.

This is not to say that malathion is harmless. In very heavy doses, far beyond those normally used in pest control, it can cause ACHE-inhibition disorders that manifest themselves in the ways described above. Aging malathion can degenerate into a more toxic compound called maloxon. Malathion's low toxic effect can be synergistically enhanced by a number of other common OP substances. If malathion is not manufactured carefully, a contaminant called iso-malathion can form during processing that greatly increases its toxicity. In 1976 a number of malathion deaths occurred in Pakistan, partly because the malathion in use at the time was contaminated with iso-malathion, but also because the U.S. Agency for International Development (AID), which had purchased the polluted malathion from an Italian firm, had neglected to make sure that the Pakistani farmers it was destined for understood its potential dangers. As a result, when they couldn't find sticks to use to mix the malathion with water, many used their bare arms, with fatal results.

The Phosvel Story

A small number of OPs have also been implicated recently as causes of another disorder, called "delayed neurotoxicity," which is unrelated to ACHE inhibition. Delayed neurotoxicity is a wasting away of the outer sheathing around nerves that can lead to paralysis. The sheathing is called myelin and its degeneration is called demyelination. A number of diseases, such as multiple sclerosis and encephalitis, are known to cause demyelination, and in the last decade we've been discovering that a few OPs apparently can also.

The most recent and potentially most hazardous case of delayed neurotoxicity, one that took place in the mid-1970s, involved Egyptian water buffaloes, laboratory hens in Chicago, and a group of workers in Bayport, Texas, employed to manufacture and pack a highly toxic OP pesticide called Phosvel, or leptophos.

The Bayport plant was owned by the Velsicol Corporation. In 1973, after successfully exporting Phosvel for a number of years, Velsicol moved to have it approved by the EPA for use in the United

States, and almost succeeded, despite the fact that toxicology data on its effects on the nervous system were very ambiguous. The story actually began in Egypt in the early 1970s, where Phosvel was heavily used to control a number of the pests that plague cotton, Egypt's foremost agricultural crop. In 1971 the hind legs of some twelve hundred water buffalo in cotton-growing areas became paralyzed, and many of the animals sickened and died. Tests conducted by the Egyptian government at the University of Alexandria that clearly implicated Phosvel were never publicly released, probably because the Egyptian economy was so dependent on cotton that the country couldn't afford to stop using it. Not until 1974, when a German toxicologist working for one of Velsicol's competitors in Egypt happened to mention the test results to an EPA scientist traveling in Finland did doubts about Phosvel's safety begin to surface in the United States.[8] When the EPA scientist returned to Washington and repeated the allegations he had heard about Phosvel, he found that a number of his colleagues had already heard much the same thing in roundabout ways from other sources.

At that time Velsicol had just formally asked the EPA to establish tolerance levels specifying the amount of Phosvel residues that could be allowed on lettuce and tomatoes imported into the United States, a preliminary step toward getting Phosvel fully registered for domestic use. In response to this request the EPA studied Velsicol's toxicology data, primarily based on tests of chickens performed by an independent laboratory in 1969, and decided that the compound looked safe. In 1974 the agency set Phosvel tolerances for lettuce and tomatoes, thus tacitly putting its stamp of approval on the product, and allowing the importation of Phosvel-treated crops from neighboring countries, especially Mexico, to commence. At that point, however, the members of the EPA's scientific staff, along with some outside scientists who had become concerned about the water buffalo deaths in Egypt, started to raise questions about the compound's safety. Learning of this groundswell of resistance to its tolerance request, Velsicol pushed even harder for approval, first insinuating that the criticism was nothing more than an attempt at industrial sabotage by its German competitor in Egypt,[9] then falling back on the tactic of requesting the EPA to convene an independent panel of experts designated by the National Academy of Sciences to review the toxicology data on Phosvel.

Simultaneously, several workers at the Bayport Velsicol plant, where the Phosvel used in Egypt and other parts of the world was made, began to show signs of bizarre nervous system disorders, variously diagnosed as multiple sclerosis, encephalitis, and encephalomyelitis (inflammation of the brain and spinal cord). In September 1974, according to *Houston City Magazine* editor Tom Curtis:

> . . . a fifth worker in the warehouse packaging unit became acutely sick. John Orville Wright was a thirty-year-old Army veteran with two tours in Viet Nam who had worked four years for another Houston chemical firm. He had happily gone to work for Velsicol early in July. The job paid better than $5 an hour and the plant was less than a ten-minute drive from his home in La Porte. After accidentally breathing some Phosvel fumes, Wright experienced a choking sensation, difficulty in breathing, profuse sweating, a tingling sensation in his arms, difficulty remembering things, and trouble judging distances. After examining him, Dr. Miller, the Velsicol physician, also detected blurred vision. Miller diagnosed the problem as allergic asthma possibly secondary to a chemical irritation and treated Wright with a tranquilizer, antihistamines, and a bronchial dilator. A few days later Wright's symptoms were worse than before, and Miller hospitalized him at Pasadena Bayshore, where he remained about five days. With his chest pains, troubled breathing, and night sweats now compounded by weakness in his legs and trouble keeping his balance, Wright went in November to see his family physician, who diagnosed Wright as having a narrowing of the urethra and prostatitis, and during a two-week hospitalization operated on him for that.
>
> In February 1975, Wright, no better, saw still another physician, Dr. Jack N. Alpert, who put Wright in the hospital again, ordered a spinal tap, and for the first time diagnosed him as having a demyelinating disease, which Alpert assumed to be caused by multiple sclerosis. By the third week of March 1975, Alpert released Wright to go back to work, but, as his wife recalls, "he was stumbling so bad they told him to take a week off and rest." He was finally fired on April 4. Mrs. Wright recalls that Jim Hacker, the personnel manager, "explained to us that it wasn't that John wasn't a good worker but that if he was to stumble and fall his safety as well as that of the other workers would be endangered." Wright says, "They gave me my walking papers in a real gentlemanly, mild way."[10]

At about this time, in late 1974, an EPA pathologist, Dr. Howard Richardson, reviewing the original Phosvel toxicology tests performed for Velsicol in 1969, discovered that critical slides of nerve tissue were unreadable and had been when the tests were first performed. In other sections of the toxicology reports, moreover, he found descriptions of the effects of Phosvel on one hen (shown below) that sounded exactly like the effects of delayed neurotoxicity:

Test Animal 161 (400 mg/kg)

TIME AFTER DOSING	OBSERVATION
0	Normal
19 hours	Comb droopy; cannot remain standing
24 hours	Barely able to walk
28 hours	Very unsteady; weak
44 hours	Comb droopy
2–13 days	Normal
SECOND DOSE	
19 hours	Reluctant to stand; legs sprawled in front
2 days	Cannot stand; no control left leg
3–5 days	Down on hocks; cannot stand or walk
6 days	Same; comb dark
7–10 days	Down on side
11 days	DEAD

Dr. Richardson immediately arranged for a duplication of the first series of tests, and soon had fresh evidence clearly showing delayed neurotoxic effects in all test hens. Using an improved tissue preparation technique that he had developed with his wife, also a pathologist, Richardson also found that microscopic examination of nerve tissue from this second test group of hens "demonstrated neural lesions in all test groups."[11]

Dr. Richardson also tried to inspect the Bayport plant but was refused admittance by Velsicol officials. Shortly afterward his work on the Phosvel case came to an untimely end when he and his wife were killed in a plane crash in Australia.

In 1976 the NAS committee set up at Velsicol's request to review the toxicological data on Phosvel recommended that the tolerances for imported lettuce and tomatoes, granted the year before, be withdrawn. Looking back, there's little doubt that many tons of lettuce and tomatoes imported from Mexico into the United States during that year were contaminated with Phosvel residues. However, it is also important to understand that, as in the case of farm workers exposed to OPs, much of what we know about Phosvel's toxicity in humans is due to an episode that primarily involved occupational exposure through inhalation and skin absorption, not ingestion. The precise relationship between acute occupational exposures and the far lower ingestion exposures of anyone in the general public who may have eaten Phosvel residues is difficult to determine.

Responding to belated reports by Velsicol of health problems among workers at its Bayport plant, NIOSH began investigating the case in 1976. Shortly afterward Velsicol withdrew its request for domestic registration of Phosvel and halted production, claiming that the market had dried up. However, it continued to export the 680,000 kilograms it had managed to stockpile in the meantime.

Phosvel was an especially hazardous compound because it was not acutely toxic enough to make a worker exposed to it feel sick enough to get away from it. Thus men in the Bayport plant were able to tolerate exposures that caused chronic damage to them. Phosvel is also one of the rare OPs that is as environmentally persistent as are many CHIs and that bioaccumulates in body tissues. The other three OPs suspected of causing delayed neurotoxicity in mammals—EPN, DEF, and merphos—are so much more acutely toxic than Phosvel that it is unlikely an individual could stand enough of an exposure to risk delayed neurotoxic effects. DEF, in addition, has a very repugnant odor. Because of the Phosvel incident and the discovery of delayed neurotoxic effects in other OPs, the EPA now requires that all new OP compounds must be extensively tested for delayed neurotoxic effects before they can be registered.

CARBAMATES

The third major category of synthetic pesticides, carbamates, is also the newest. Although the first carbamates were synthesized as early as 1935, the advent of the carbamate era in the United States did not occur until eighteen years later, when Dr. Joseph Lambrech of the Union Carbide Company created what was to become the most widely used carbamate in the world today, carbaryl, also known as Sevin. In 1959 carbaryl had just been approved for marketing in the United States; a little over a decade later this country was producing 25 million kilograms per year.*

*As of 1974, carbamates and organophosphates combined accounted for about 35 percent of the total amount of pesticides used in the United States. Chlorinated hy-

The derivation of carbamates can be traced back to use of the only known natural carbamate plant, the calabar bean, in African rituals of trial by ordeal. In Nigeria accused wrongdoers were forced to drink a milky potion containing ground-up calabar beans. "Either the prisoner was fortunate enough to regurgitate the 'milk,' in which case he was declared innocent and allowed to go free, or he would develop the shakes, froth at the mouth, and quickly die, thus establishing his guilt."[12]

Carbamates vary widely in persistence. Carbaryl biodegrades very rapidly in the environment after application, almost disappearing within days. Carbamate toxicities also vary widely: Carbaryl's acute oral LD_{50} in rats is about 500 mg/kg, whereas carbofuran, another popular carbamate insecticide, has an acute oral LD_{50} of 11 mg/kg in rats. Carbamates are not stored in body fat tissues, so the small residues that do remain in food ingested by man are rapidly excreted.

Like OPs, carbamate toxicity is caused by ACHE inhibition. However, unlike OPs, the ACHE inhibition caused by carbamate compounds is rapidly reversible. A laboratory animal given enough carbamate to suffer moderately severe signs of carbamate intoxication—heavy salivation, muscle tremors, uncontrollable urination and defecation—recovers within hours.

ALTERNATIVE INSECTICIDE METHODS

Over the last few decades, in response to mounting concern about the terrestrial environment and the well-being of the species that occupy it, man's endlessly inventive mind has come up with numerous alternatives to the use of massive applications of toxic chemicals. Some of these alternatives have already been in routine use for years. Others are still highly experimental and may never become truly practical for a variety of reasons—because they are too expensive, because they waste energy, because they are too specific, killing only one type of pest species, or simply because we don't know enough about them yet to use them safely and effectively, and further research on them appears to be too slow and too expensive to make it worthwhile.

---

drocarbon compounds accounted for 24 percent, and that figure has certainly declined since, due to the increasingly stringent restrictions placed on every type of CHI pesticide.

Some methods of alternative pest control involve only the simplest changes in the way crops are planted. Allowing less space between plants, planting a secondary crop between the rows of a primary crop, using more water or less, or even just making sure that a crop is properly and adequately fertilized—all these are basic techniques that have been used on occasion with some success in the control of insect pests. Rotating crops from one year to the next is another simple pest-control technique that works by periodically depriving a pest species of its main source of nutrition, thus reducing its population. Even as simple a step as planting after a pest species has already hatched and run its life cycle has proven effective in some instances.

Another well-established alternative pest control method springs from the realization that insects get diseases, too. Bacteria, viruses, fungi, and other microbes that prey on pest insects have in numerous cases been successfully used by farmers to protect their crops. One well-known example is the use of a particular bacteria—"milky disease" spores—that prey on Japanese beetles. In this process beetle grubs are cultured or collected, deliberately infected with the bacteria, and stored. Then, just before use, they are ground up into a powder, which is applied to the area needing protection. The bacterial spores in the powder are extremely persistent and may continue to reinfect and kill new beetle grubs for as long as twenty years. Unfortunately, there is the inevitable drawback to this approach: The powder is hard to make and distribute in sufficiently large quantities.

Another well-established alternative is based on the phenomenon of resistance. Most of us are aware that insects can and do gradually evolve into new strains that are no longer susceptible to the doses of a pesticide that would have easily killed their progenitors. In these cases the only alternative has been to increase the dose of the same pesticide, risking environmental damage and threats to other species, or switching to a different, often untried pesticide altogether.

However, relatively few people understand that resistance occurs in plants, too, and that this phenomenon can be put to work for mankind. Resistance occurs naturally among many plant species, but usually food and fiber crops must be bred to become resistant to a specific pest species, and this has traditionally been a long, slow process, but one which has in many cases yielded strikingly effective results. Cotton, potatoes, tobacco, wheat, oats, tomatoes, beans, and alfalfa are just some of the crops that have been successfully bred to become resistant to specific pest species in the past.

This is an area in which a great deal of exciting new work has been done in recent years, stimulated by rapid advances in the technology known as genetic engineering. Today, the real promise of research into recombinant DNA as it applies to food production is that it may in the near future allow us to design completely new crop strains, rather than having to breed them from old ones. The resulting crops will not only be resistant to pests, but generally hardier, stronger, and more adaptable to a wide variety of climatological and geographical conditions than their predecessors. As of this writing, genetic engineering has brought us closest to a breakthrough in the area of redesigning yeasts that are better suited to the fermentation of alcohol fuels from plant matter, but food crop engineering is not far behind.

Another major area of alternative pest control depends on interfering with insect reproductive cycles. One typical version of this method is to sterilize—by radiation or chemical means—large numbers of one sex of a pest species and then release them to mate. Because of the presence of large numbers of sterilized members of the pest population, fewer offspring are produced, and the species' numbers dwindle. (This technique can backfire, as California learned during the Medfly debacle, if the "sterile" insects released are not in fact sterile.) Another approach is to use the chemical substances secreted by one sex to attract swarms of the opposite sex to a trap that can then be baited with a chemical poison. This method has been used successfully against gypsy moths in the Northeast, using a substance known as "gyplure," and another variant was recently tested by researchers at the University of California at Davis. As part of a campaign to eradicate the artichoke plume moth, the Davis researchers spread a synthetic female sex attractant in an artichoke field. Its presence caused males to seek the females, but when they couldn't find them they became disoriented and did not mate at all, causing a significant decrease in the moth population in that area.

Yet another approach in this area, one that has just begun to look promising in recent years, involves the use of insect hormones to speed up or delay pest maturation into normal adulthood. If the pests are of a type that is most damaging as adults, such as mosquitoes, then juvenile hormones can be used to keep them immature—and harmless—until they die. If, like most species, they are most damaging as a juvenile larvae, the use of adult or "antijuvenile" hormones will speed their passage through the larval stage, reducing the amount of time they have to cause crop damage.

There are also a number of physical insect repellent systems, ranging from the use of ultraviolet light to attract insects to an electrocution grid, to infrared light systems used to kill bugs on grain conveyor belts, to experimental X-ray, gamma ray, and electromagnetic wave systems, which kill by heat, and even some sonic and ultrasound systems.[13]

IPM

For a good two decades now, some entomologists and pesticide researchers have been advocating and some farmers have been following a remarkably level-headed approach to pest control that borrows elements from many alternative methods, including some of those described above, and merges them into a flexible strategy known as Integrated Pest Management, or IPM. There is no such thing as a "typical" IPM program, since pest problems vary from region to region, but a hypothetical IPM program might involve use of a resistant crop strain, natural pest predators, delayed planting, and a synthetic sex attractant that collects insects in one place and then kills them with a chemical bait. A common characteristic among all IPM methods is a strong reluctance to use large applications of chemicals except as a last resort. Small amounts of chemicals are a welcome part of many IPM programs, but the conventional approach of heavy reliance upon chemicals and nothing else is avoided.

Unfortunately, environmental interdependence currently tends to work against IPM farmers. When their neighbors use spray, pests leave their fields and overwhelm the less aggressive methods that IPMers use, which means that they are then forced to spray also. This wipes out their populations of natural pest predators, and they have to start all over again. Fortunately, new bug populations are less expensive than chemicals.

PYRETHRUM: NATURE'S INSECTICIDE

According to a leading authority on pyrethrum and its derivatives, John E. Casida of the University of California at Berkeley,[14] pyrethrum was known and used in Europe at least a century ago and in the Middle East hundreds of years before that. Some of the earliest references to pyrethrum suggest that the Crusaders may have learned to use the powder of the dried pyrethrum flowers to control body lice during their forays against the infidel.

Until World War I the world's pyrethrum supply came primarily from the region known as Dalmatia, now part of Yugoslavia. Capitalizing on the disruption of trade in that part of the world caused by the First World War, Japan became the dominant exporter of pyrethrum flowers after 1915, and by 1935 was raising an annual crop of some 12.6 million kilograms on nearly 72 million acres of land. During the 1930s, however, superior pyrethrum flowers grown in Kenya began to attract the attention of pyrethrum buyers, and

Pyrethrum Plant

Figure 21

African crops soon were offering stiff competition to the Japanese product. When the Second World War closed off pyrethrum shipments from Asia, Kenya moved into place as the foremost pyrethrum growing and exporting nation in the world, a position it still holds today. Shortly afterward the advent of DDT and other synthetic organic pesticides dealt the fledgling Kenya pyrethrum industry a near-fatal blow. In three years, from 1945 to 1948, production plummeted from 5.6 million to 1.4 million kilograms. But as information about the possible health and environmental hazards of DDT and other synthetic pesticides began to surface in the 1950s, demand slowly grew again. Today, pyrethrum is also grown in other parts of Africa, Brazil, Ecuador, and Japan.

The insecticidal part of the plant (known as pyrethrin) is concentrated in the flower. Like all other classes of major insecticides in use today, pyrethrins are nerve poisons, but the exact mechanism of their effect is still not fully understood. Lethal though they are to a wide range of insects, the toxicity of pyrethrins to mammals is extremely low. According to Dr. Casida, pure pyrethrin's oral acute LD_{50} in rats is somewhere between 400 and 800 mg/kg. In chronic toxicology tests rats have been fed up to 250 mg/kg for as long as two years with no adverse effects. A study by the World Health Organization in 1970 in which pregnant rabbits were fed 90 mg/kg a day revealed no teratogenic effects at all.[15] There are very few substantiated cases of human poisoning from pyrethrum. The last one on record occurred in 1889 when a two-year-old child died after swallowing half an ounce of pyrethrum powder, a dose equal to about 1,000 mg/kg in a typical two-year-old weighing 13.5 kilograms (30 pounds). One pyrethrin researcher reported swallowing 50 mg of pyrethrin with no effect at all, other than a numbness in his tongue and lips.[16] Not too many decades ago pyrethrin tablets were commonly ingested for tapeworm control; a typical daily dose was 10 to 20 milligrams. Recent studies of the chronic effects of pyrethrin on mammals show no evidence of their being mutagenic, teratogenic, or carcinogenic.[17] According to Casida, the only organ that pyrethrins may damage in mammals is the liver, and signs of this damage appear only at high doses. The only regularly adverse health effects of pyrethrum exposure is an allergic response in some pyrethrum workers, who experience itching, reddening, swelling, and cracking of the skin, a condition that is often due to other impurities in pyrethrum mixtures rather than the pyrethrins themselves, and usually clears up when the worker is removed from the contaminated environment.

Pyrethrins are also highly biodegradable. They are not harmful to farm animals, pets, or birds, but are quite toxic to fish. They have a long history of safe and effective use on all kinds of plants, including most food crops, under a wide variety of conditions, and against a broad spectrum of insect pests, including flies and mosquitoes. During the many decades they have been in regular use, there has been little or no development of insect resistance against pyrethrins. They are still commonly used today in household aerosol sprays, on international aircraft flights, on stored foods, and for certain industrial purposes.

The major argument against pyrethrum in the past has been economic. Since pyrethrins are so highly biodegradable, more applications are necessary than with persistent synthetic insecticides, and this of course increases their cost. One minor benefit of the recent escalation in oil prices is that by making substances like synthetic insecticides, which are derived from petroleum, more expensive, it has given a boost to less toxic alternatives, such as pyrethrins.

In recent decades the development of highly specific low-volume application methods for pyrethrins and the use of synergizers that reduce the amount needed for an effective insect kill by magnifying the toxic effect of a small amount many times have also helped to make pyrethrum insecticides more competitive. Today, the demand for pyrethrum-based insecticides is about double the supply. Much of the difference is being made up by synthetic pyrethrins, known as pyrethroids, though these have certain drawbacks.

PYRETHROIDS

The theoretical basis for chemical synthesis of pyrethroids has been known since early in this century, but very little application of that knowledge to the development of practical pyrethroids took place until the late 1960s. Until 1963 only one synthetic pyrethroid had been brought to full commercial viability. Since then, however, researchers in Japan, England, and the United States have added some half-dozen pyrethroids to the list of those in active use, and the search for new and ever more effective synthetic analogues of the pyrethrum plant continues today.

However, as is always the case, we are finding that our efforts to improve on nature have introduced a whole new set of complications and drawbacks. In order to make pyrethroids more effective, a great deal of effort has gone into experimenting with ways to make them more persistent, which would also make their use less expensive since fewer applications would be required. Some of the most recent pyrethroids rival DDT in their environmental persistence. However, some of the chemical adjustments that make pyrethroids more stable also make them much more toxic to mammals, so we are once again faced with a trade-off. Furthermore, researchers are finding that many insect strains, often the same ones that have become resistant to DDT, are now beginning to show signs of strong resistance to pyrethroids as well.

HERBICIDES

When we think of pesticides we usually think of substances that kill bugs. Since 1967, however, the real growth area in the pesticide field has been in synthetic weed-killers, or herbicides. In that year herbicide sales exceeded insecticide sales in the United States for the first time[18] and the gap between them has widened ever since.

In 1958 the United States produced about 27 million kilograms of herbicides. One decade later that amount had risen to over 146 million kilograms.[19] In 1974 we were using more than that for our *domestic* needs alone. By 1968, of the 407 million kg of pesticides used in the United States, herbicides represented the largest single percentage, some 45 percent, according to the EPA.[20] By comparison, insect control compounds represented only 30 percent of that total. Herbicide sales are currently growing at about twice the rate of insecticide sales.

From the viewpoint of the pesticide manufacturers this is an especially welcome trend. Despite their enormous profits the pesticide business is becoming a tougher one to make a profit in. Rising production costs, the technological difficulty of synthesizing a new compound with unique pesticidal properties, and increasingly stringent environmental regulations are making it harder and harder to find a compound that can pass all the required health and environmental safety tests and move into the marketplace. One pesticide researcher recently noted that whereas in 1956 one in every eighteen hundred new substances devised in the lab made it through screening and evaluation hurdles into the marketplace, by 1977 that rate had dropped to something like one in every twelve thousand,[21] and was growing worse all the time.

There are some very good reasons for the soaring popularity of herbicides. First of all they offer a way to escape from the drudgery of manual clearing, thinning, and weeding, and the energy-intensive work of mechanized tilling. A farmer who uses 2,4-D on his corn can grow acre upon acre, row upon row, without ever having to pull a single weed. And since weeds *do* compete with food crops for sun, water, space, and soil nutrients, it is a distinct advantage to be able to get rid of them economically and effectively. As well as being widely used in food crops, herbicides are also commonly used to help keep highway shoulders, railroad tracks, waterways, open rangeland, golf courses, parks, industrial sites, forests, and numerous other types of spaces free of obnoxious or just plain unwanted plant growth.

The boom in herbicide use isn't likely to slow down. We are rapidly approaching the day, probably less than twenty years hence, when the earth's population tops the 6 billion figure. At the same time, however, its total arable land will, according to the Council on Environmental Quality, have increased by only 4 percent. The council estimates that food production will increase by 90 percent during the next two decades, but "in the early 1970s one hectare of arable land supported an average of 2.6 persons; by 2000 one hectare will have to support 4 persons."[22]

What this means is that increases in food production, which will be absolutely essential if we are to avoid mass famines in the developing nations, will have to be produced by increasing our yields from land already under cultivation rather than by bringing new acreage under the plow. This in turn means that we will be depending heavily throughout the world on all kinds of pesticides, herbicides in particular. Although few people realize it, weeds may be a greater agricultural problem than insects. In the decade from 1950 to 1960 crop loss from weeds amounted to $5 billion, more than the cost of insect damage and plant disease combined.[23] As of 1973 the United States was producing over 150 different kinds of herbicides, and our understanding of how they work and how to make them more selective was becoming more precise every year.

Unfortunately, as one must expect of any substance expressly designed to kill living organisms, herbicides can be toxic to mammals. In general, however, their acute toxicity is much lower than that of most insecticides, partly because there are fewer similarities between plant and mammalian metabolic processes than there are between mammals and insects. But, in the last few years clear indications have appeared that some herbicides and contaminants introduced into them during their manufacture may pose significant chronic health threats to man, including cancer.

PARAQUAT

Paraquat achieved a small degree of notoriety a few years ago because it was being sprayed on marijuana crops in Mexico in an effort to cut off one source of marijuana entering the United States. The strategy failed because once they are cut and dried, sprayed plants are indistinguishable from unsprayed ones—Mexican growers simply went ahead and harvested their sprayed crops as usual and shipped them out rather than lose a sale. When this fact became known in the United States, it created a justifiable but somewhat misinformed panic in dope-smoking circles.

When ignited and inhaled, paraquat byproducts can cause a mild, transient burn in the lining of the throat. However, as numerous accidental and suicidal ingestions of paraquat have demonstrated, this herbicide is much more hazardous when swallowed. Ingestion of a relatively small amount of paraquat is followed by a deceptive period of some three to four weeks in which no symptoms are experienced. But within about one month the first signs of irreversible and fatal pulmonary fibrosis, a type of scarring of the lung tissue that obstructs breathing, begin to appear. Although most of what we know about paraquat's toxicity has been learned from cases in which more than 70 cc have been ingested, as little as 15 cc (about one tablespoon) is a lethal dose.

Paraquat is widely used in the United States as a preemergent (before the crop comes up) spray to control weeds. It cannot be used once the crop has appeared because it is nonselective and will kill desirable as well as undesirable plants.

On July 12, 1981, the Federal Drug Enforcement Administration announced that it was working closely with state governments in California, Florida, Georgia, North Carolina, and South Carolina in planning a possible paraquat spraying program to suppress domestic marijuana production. Some states, concerned about paraquat's acute toxicity, have placed special restrictions on its use, but as of this writing the EPA has not.

2,4-D AND 2,4,5-T

Both 2,4-D and 2,4,5-T are members of a chemical family known as "chlorophenoxy" herbicides. Chlorophenoxies are systemic herbicides, which means that if they are sprayed on any part of a plant, or even the ground around it, they will quickly spread throughout it (a process known as "translocation") and kill it. Chlorophenoxies cause a host of interrelated, somewhat mysterious effects on plants, often likened to a type of plant cancer. Photosynthesis slows, some parts of the plant swell or elongate, others, like leaves, stop growing entirely. Roots quit taking in water and soil nutrients; the tissues that distribute water and nutrients throughout the plant become blocked and die.

First marketed in 1944, 2,4-D is one of the oldest herbicides around, and one of the most widely used. Experts estimate that 75 percent of all weed control of any kind involves just three basic compounds, one of which is 2,4-D. It is used in some fifteen hundred different products, including many specifically formulated for home use. It is highly biodegradable, lasting at most from a few weeks on sprayed vegetables to a few months in water. In the human body it is not well absorbed through the skin but passes readily through the walls of the intestine, and is eliminated within days.

The acute oral LD_{50} of 2,4-D in rats is about 275 mg/kg, which makes it a moderately toxic substance. Studies of its effects on poisoned animals have found evidence that it damages the liver and kidneys, and under some conditions can cause disorders of the central nervous system, including paralysis. It is a teratogen in rats, mice, hamsters, and pigs at fairly high dosage levels. It does not appear to be mutagenic, though on this issue there is ambiguity in the data. There is clearer evidence that 2,4-D is an animal carcinogen, but again the data are not entirely consistent on this point, either.

Pesticide applicators acutely exposed to 2,4-D have reported symptoms of headache, loss of appetite, dizziness, weakness, gastritis, vomiting, chest pain, and blackouts. Chronic exposure seems to be linked to appetite loss, loss of taste and smell, headaches, fatigue, and nonspecific abdominal pain. In a very few cases signs of degeneration of peripheral nerves have been reported, and in one case an autopsy of a fatality caused by 2,4-D poisoning disclosed extensive damage to the outer sheathing of the nerves in all parts of the brain. Experts believe the lethal dose of 2,4-D in humans is between 80 and 800 mg/kg body weight, making it significantly less acutely toxic than parathion, but quite a bit more toxic than malathion. The EPA is currently developing better information on 2,4-D's toxicity and reviewing its registration to see if additional restrictions on its use are advisable.

2,4,5-T, a close chemical cousin to 2,4-D, appears to be quite a bit more hazardous. What we know about it to date indicates that it is not carcinogenic or mutagenic, and is only teratogenic in laboratory animals at near lethal doses. The acute oral LD_{50} in rats is about 500 mg/kg, in mice and guinea pigs it is 380 mg/kg, and in dogs it is 100 mg/kg. Acutely toxic doses appear to damage the heart, bone-marrow, kidneys, and the lymph system in animals.

However, the toxicity of 2,4-5-T has historically been difficult to isolate and study because the way it has been manufactured has in the past invariably contaminated it with an extremely toxic compound known as TCDD, a type of dioxin. Most of the poisoning incidents ostensibly involving 2,4,5-T are more likely incidents involving TCDD poisoning.

TCDD

Studies have shown that TCDD has an acute oral LD_{50} of about 100 μg/kg in rats. It is at least one hundred times as acutely toxic as parathion. Autopsies of lab animals reveal that TCDD causes extensive damage to the liver, kidneys, heart, thyroid, lymph system, urinary bladder, and gastrointestinal tract. It also causes suppression of the immune system, making animals more susceptible to infection. TCDD is clearly carcinogenic, mutagenic, and teratogenic in laboratory animals.

In human beings exposed to TCDD the most common effect is chloracne. Other signs of TCDD exposure in man include inflamed eyes, liver damage, kidney problems, gastrointestinal disturbances, loss of appetite, weight loss, fatigue, decreased resistance to infections, especially of the respiratory tract, loss of smell, taste, and hearing, tissue hemorrhaging, nerve inflammation, heart disturbances, neuromuscular disorders, and psychological changes.[24] Its ability to cause cancer, mutations, and birth defects in humans is still highly debatable.

TCDD has no telltale warning properties—it is invisible and odorless. It is readily biodegraded by sunlight, but may persist for years in the soil or in other places hidden from the sun. The only certain way of destroying TCDD is to incinerate it at temperatures higher than 1000°C in special incinerators.

To date, what little direct information we have about the effect of TCDD in humans exposed to it has come from occurrences in which the major routes of exposure were inhalation and skin absorption. TCDD residues are almost never detected in foods, and on the rare occasions that they have been, the levels were so low as to be insignificant.

There have also been a number of industrial accidents in which workers were exposed to TCDD, with ambiguous effects on their health. To date there have also been only two major public exposures to TCDD-contaminated substances, one at Seveso, Italy, in 1976, and the exposures of Vietnamese and American civilians and servicemen to Agent Orange during the Vietnam War. In neither of these cases is it yet clear what health effects the exposures have caused. Barring the possibility of industrial accidents such as the one at Seveso, virtually the only conceivable exposure to TCDD on the part of the public today is to badly made 2,4,5-T, and it is precisely for this reason that the manufacture of 2,4,5-T is rigidly controlled.

Seveso

Shortly after twelve thirty P.M. on Saturday, July 10, 1976, the residents of the small town of Seveso, Italy, were startled by a shrill whistling sound issuing from the local Icmesa-Givaudan chemical plant, located in the neighboring town of Meda. Moments later a grayish-white plume could be seen rising some fifty meters into the air and drifting slowly toward the southeast, downwind of the plant. As it passed overhead, a fine shower of minute particles rained down on everything in its path—houses, trees, dogs, cats, chickens, rabbits, and people.

Seveso is a quiet town of seventeen thousand residents, only twenty kilometers from Milan, in the north of Italy. It is situated in a lush region of Lombardy, once renowned for its rich land and its skilled artisans. More recently, along with much of the area around Milan, it has become increasingly industrialized. The Icmesa plant is a subsidiary of the giant Swiss chemical firm Hoffman-La Roche. In 1976 one of the plant's major products was a substance known as TCP (sodium trichlorophenate), a basic ingredient used in making 2,4,5-T, among other things.

Workers at the Icmesa plant soon discovered that the shrill noise was caused by superheated chemicals escaping from a vessel in which TCP was being made. When for some unknown reason the mixture became dangerously hot, a safety vent ruptured and the vessel spewed its contents into the atmosphere. Later estimates placed the temperature inside this vessel at 400°C or more.

Samples of soil and vegetation near the plant were collected and rushed to Hoffman-La Roche headquarters in Basel, Switzerland, for analysis. Plant officials recommended that local residents avoid eating fruits and vegetables exposed to the escaped chemicals until they could be evaluated, but emphasized their belief that there was no serious danger.

According to an account of the Seveso incident written shortly afterward by John G. Fuller,[25] almost as soon as the cloud had passed overhead, many Seveso residents started to experience headaches and sore eyes, probably caused by a caustic alkaline substance used in the manufacture of TCP. But, reassured that there was probably no danger, they continued with their weekend routines. Some heeded the ban on consuming fruits and vegetables. Many did not, believing that washing or boiling them would remove whatever had contaminated them. Most also continued during the next few days to consume the rabbits and chickens—all fed on local vegetation—that Seveso citizens commonly raised for the table.

Within a few days it became apparent that the incident was not soon going to pass from memory. Many adult Sevesans were feeling sicker with each passing day, complaining of backaches, stomach pain, liver and kidney pain, and nausea. Some adults and many children who had been outside at the time of the explosion began to develop an ugly, itchy rash on their arms, legs, and faces. In many cases the rash grew steadily worse, turning into large running blisters and burnlike sores all over the body. At about the same time, residents began to see birds dropping out of the sky, and found their household animals dying all around them.

Not until July 20, ten days after the accident, did Givaudan officials notify Lombardy health officers that the heat of the vessel had caused the formation of TCDD, which had been found in high concentrations in many of the samples taken from the area around the plant. Later, it was estimated that at least three kilograms of TCDD, had been released over some seven hundred acres of Seveso and neighboring areas. In the areas of heaviest contamination, immediately downwind from the plant, TCDD concentrations in the soil averaged 580 $\mu g/m^2$ and ranged from zero to as high as 5,477 $\mu g/m^2$.

On July 26, Seveso's mayor ordered the evacuation of some 179 people from the most contaminated zone. A few days later another 700 people were ordered to leave their homes, taking only the clothes on their backs and one suitcase with them. In many cases, Fuller notes, a squad of armed police had to make sure that the order was obeyed. Evacuees were housed in nearby hotels at the expense of Hoffman-La Roche.

The entire area of highest contamination, labeled Zone A, was sealed off with barbed wire and patrolled by armed guards. The few technicians allowed to enter to continue sampling were required to wear face masks and protective clothing. They found that almost every animal in Zone A was dead or dying.

The first autopsies of animal victims of the accident revealed extensive damage to internal organs: "The subcutaneous tissues were abnormally swollen. The fatty tissues of the animal were totally degenerated. The trachea had massively hemorrhaged. The liver and kidneys were overwhelmed with lesions."[26]

To prevent the spread of contaminants, eggs, cheese, and milk within Zone A and bees and beehives as far as three miles away were destroyed. But nothing could be done about wild animals and birds entering from outside the area, or the dust carried by the wind, by cars on the nearby autostrada to Milan, and even by foot traffic around fenced-off areas. (Some six months later vegetables in Milan markets were found to be contaminated with six hundred times the maximum allowable level of TCDD.)

Meanwhile in Seveso the process of evaluating the extent and significance of the accident continued. More samples were taken of local soil, water, vegetation, and air. Scrapings and wipings from walls and other surfaces, both inside and outside buildings in the contaminated area, were obtained. Trenches were dug and samples taken at various levels below the soil surface to determine how deep the TCDD had settled. The results showed significant penetration into the soil, exceeding 1,000 $\mu g/m^2$ in many places.[27]

By February 1977 signs of chloracne had appeared in some four hundred Sevesans, mostly children. An intensive medical monitoring program of the eight thousand individuals considered to be at highest risk had been initiated. One Icmesa worker had died of cancer of the liver. An autopsy of one Sevesan woman who died of cancer of the pancreas indicated that her body contained a total of some 40 μg of TCDD.[28] Three of six children born to Seveso women who were pregnant at the time of the accident had birth defects: Two had malformed intestinal tracts and one a cleft palate. In defiance of the Church many Seveso women, fearing they would bear deformed children, had abortions.

Hoffman-La Roche continued to pay for the expenses of the evacuees, as well as the wages of the employees at the idled plant. It also offered to reimburse local farmers for the loss of their land and crops. The Italian government allocated $48 million to help defray the mounting costs of the accident, and Hoffman-La Roche contributed $11.5 million, but even these amounts were not sufficient to cover the rising expenses. Questions about why there had been no backup system to cool down overheating reactors, no containment system to keep potentially hazardous chemicals from escaping into the environment at large, and why it had taken a full ten days for Hoffman-La Roche to report the presence of high levels of TCDD in environmental samples to local authorities went unanswered.

In late 1976 reclamation operations began at Seveso. Sunlight causes TCDD to break down rapidly into harmless byproducts where it lies exposed in vegetation or on the earth's surface. Below the soil surface, however, or in tiny porous cracks in plaster and masonry, it can last for years. About twenty-five centimeters (ten inches) of topsoil was removed throughout the most-contaminated areas and replaced with fresh topsoil. Tests later showed that this removed about 90 percent of the residual TCDD in the ground. Walls were scraped or scrubbed and then painted or varnished. Absorbent household materials—linoleum, wallpaper, furniture—were simply thrown out.

By the end of 1976 TCDD levels in the soil at Seveso had declined by 50 percent. However, there is still TCDD in the soil at Seveso today, and because it is in the soil it is in the dust, the air, and therefore in the lungs of anyone who lives or visits there.

Despite the unprecedented exposure of such a large number of people to such high levels of TCDD, the health effects of the Seveso accident have to date been remarkably light. All those with severe chloracne have recovered. In only two cases were there any residual scars and these are rapidly clearing up.[29] Extensive medical monitoring has so far revealed no significant abnormalities in the high-risk group of exposed citizens. It is not clear whether the three cases of birth defects reported in January 1977 were due to TCDD exposure or to vastly intensified medical surveillance and reporting procedures that were initiated following the accident.

However, it has only been six years since the Seveso accident, and it will be another five to ten more before unusual increases in cancer and other chronic diseases with long latency periods begin to show up in the high-risk groups of exposed Seveso citizens. Not until the mid-1980s, at the earliest, will we learn the true consequences of one of the worst public exposures to TCDD in history.

AGENT ORANGE

Concern about the toxicity of 2,4,5-T first arose during the Vietnam War, where its use as one ingredient in a defoliant known as Agent Orange exposed thousands of servicemen and civilians to its effects. Some forty-five thousand metric tons of Agent Orange were used in Vietnam. Unfortunately the 2,4,5-T used in Agent Orange was poorly made and as a consequence was more highly contaminated with TCDD than any 2,4,5-T made since. Experts estimate that its TCDD concentration ranged from 1 to 50 ppm and that some fifty kilograms in all of TCDD were dispersed into the environment in Vietnam. By 1970 claims that exposure to Agent Orange was causing large increases in liver cancer and birth defects in Vietnam were appearing. To date these claims have not been scientifically substantiated.

Veterans returning to the United States from the war, particularly the "ranch hands" who had been most heavily exposed to Agent Orange and other herbicides, soon began to deluge the Veterans Administration with complaints of skin disorders, cancer, birth defects in their children, and impotence. Both the VA and the Pentagon turned a deaf ear to these complaints until June 16, 1981, when Congress finally passed a bill ordering the VA to treat them. Whether the VA will in fact comply with this order remains to be seen. Furthermore, the recent addition of such prestigious voices as that of the National Academy of Sciences (NAS) to the chorus of those urging a thorough investigation of the health consequences of Agent Orange on Vietnam veterans has finally forced the Pentagon to respond to the issue by deciding to conduct a formal study of these men. Unfortunately, though NAS and other groups advised that a civilian research group be retained to conduct the study, the Pentagon has assigned it to the Air Force, raising doubts in many minds about the study's credibility before it has even begun.

Regardless of this study's conclusions, we probably won't know for another decade, when chronic effects begin to appear in exposed individuals, just what the full impact of Agent Orange was. A total of some 4 million veterans were possibly exposed to Agent Orange—it could turn out to be the asbestos story of the Vietnam era. In the meantime lawsuits have been filed by lawyers representing some seven thousand veterans against Dow, Monsanto, Hooker, and a number of other chemical companies involved in the manufacture of Agent Orange.

The military stopped using Agent Orange in Vietnam in 1970, and some uses of 2,4,5-T in the United States were suspended at about the same time, but most were allowed to continue. A maximum allowable TCDD content of 0.1 mg/kg was established for 2,4,5-T; most formulations of 2,4,5-T used today contain less than 0.025 ppm, a quarter of that level.

However, the controversy over 2,4,5-T has by no means died

down, perhaps because of its indelible association in the public's mind with a war that we have very mixed feelings about. In 1978 seven women living in a rural area in Oregon, near the town of Alsea, an area that had been sprayed by the U.S. Forest Service with 2,4,5-T a few months earlier as part of a program to control the growth of alder trees, claimed in a letter to EPA Administrator Douglas Costle that they believed the spraying was linked to an unusually high number of spontaneous abortions among them. After conducting a brief study, known as Alsea II, that seemed to support this contention, the EPA, acting under extreme public pressure, issued an emergency suspension of nearly all uses of 2,4,5-T on February 28, 1979, exempting only economically critical applications on rangeland and in rice fields.

The suspension turned out to be scientifically unjustifiable. Numerous subsequent reviews of Alsea II's data and methodology soon thoroughly discredited it, making it clear that there is no evidence whatsoever of an unusually high number of spontaneous abortions in the Alsea area during the period in question. The suspension, however, has remained in effect, and efforts to evaluate the true hazards of 2,4,5-T more thoroughly continue today.

The most credible evidence to date that chlorophenoxy herbicides are chronically as well as acutely toxic has come from a series of studies conducted in Sweden since 1974[30] focusing on individuals who are occupationally exposed to chlorophenoxies in the course of applying them on farms, in forests, and along railroad tracks. These studies have produced strong, scientifically sound evidence that 2,4,5-T causes a 600 percent increase in the risk of developing certain types of cancer, known as "soft-tissue sarcomas." However, it is still unclear what levels of 2,4,5-T are hazardous, and whether it is exposure to 2,4,5-T alone or in conjunction with other herbicides that causes this sharply increased risk.

Furthermore, despite this strong evidence that certain chlorophenoxy herbicides are hazardous for those who apply them, there is still no indication that the general public shares that risk. The exposure of an applicator to any pesticide is invariably much higher than for the average member of the public, whose exposure to pesticides results primarily from the use of products formulated for home use, which are far less potent than any other kind of formulation (though still hazardous if not used according to instructions), through drift from aerial spraying, which is now far less common than it was a few decades ago, precisely because of the public exposure problems it causes, and from residues in foods, a topic we examine more closely next.

PESTICIDE REGULATION

The major federal tool for regulating pesticide sale and use is FIFRA, the Federal Insecticide, Fungicide, and Rodenticide Act (Public Law 92-516), which became law on October 21, 1972. FIFRA is not the first federal attempt to regulate pesticides—as early as the late 1940s legislation had been enacted granting the USDA some authority in this area, but its powers were limited to matters involving interstate commerce. Within each state, pesticide regulation was left in the hands of local and state officials. As it had in other areas of environmental protection, this approach produced a patchwork of widely varying responses to the problems associated with pesticide use. It wasn't until after the job of pesticide regulation passed from the hands of the USDA to the EPA in 1970 that we began to develop national pesticide control laws with real teeth in them.

Like other environmental protection laws and its own legislative predecessors, FIFRA embodies a licensing approach to pesticide regulation. It requires that all pesticides used in this country be registered with the EPA, which assigns different classifications to them based on their toxicity, stipulates labeling and application conditions, and, most important, requires that manufacturers prove their products are safe, by conducting extensive bioassays and other kinds of tests before registration. FIFRA's extensive requirements for health and environmental impact testing were its major effect on the pesticide industry, which consistently argues that, since testing has become so expensive and time-consuming, these requirements have resulted in the suppression of the development of new pesticides.

Pesticide laws existing before 1972 required some toxicology testing, but it was mainly oriented toward acute rather than chronic effects. Because most pesticides in common use today were registered before FIFRA came into effect, we don't always know all that we should about them, particularly about their chronic effects. The EPA was directed to reevaluate these older registrations in 1972, and can require a pesticide manufacturer to conduct new tests on a registered compound if it perceives a need for them. However, there are so many of these older compounds in use that this process is a very slow one, and will not be completed until the end of the century at the earliest. Until then we won't have all the information we need to be certain that all the pesticides we're using today are safe.

The EPA's major enforcement weapons are cancellation and suspension of a pesticide's registration. (Technically there is no such thing as "banning" a pesticide.) Before a pesticide is registered, the burden of proving its safety falls on the manufacturer. But once it has been registered, the burden of proving that it may not be safe shifts to the EPA. That is, if questions about its safety are raised after it has been registered, it is up to the EPA to prove its case.

Questions about the safety of registered pesticides can and do arise at any time—the fact that they have gone through the registration procedure does not automatically render them immune from further investigation. The administrative tool for debating these questions is something known as the RPAR process, which stands for "rebuttable presumption against registration," and simply means that the EPA has received or uncovered information that raises doubts about the compound's suitability for continued registration, and is inviting interested parties to rebut or comment in any way on this information.

For example, the EPA may learn that a pesticide widely used on corn appears to be killing fish in streams near areas where it is used. Predators are eating the fish and dying, and there is some evidence that the compound may be entering the food chain. After checking to make sure the reports are reliable, the EPA will in essence invite discussion of this issue by publishing an RPAR notice in the *Federal Register*. It is then up to any interested party—manufacturers, the USDA, farmers who use the pesticide, governments of corn-growing states, fishermen, environmental protection groups, and so on—to rebut or support the RPAR allegations within a certain amount of time.

The EPA reviews these responses, carefully compares the risks and benefits of use of the compound, and comes up with a proposal. If there is a way to alleviate the risks while still allowing the benefits to continue, the proposal will suggest this, usually by recommending changes in the way the compound is formulated, labeled, or used. For example, in the case above the EPA may have learned that the compound breaks down completely over a distance of five miles from its point of application, so it can propose that five-mile-wide buffer zones be established around all local waterways in corn-growing areas where this pesticide is used. This proposal is then reviewed by a scientific advisory panel, and is published in the *Federal Register*. If no significant objections arise at this stage, it is then implemented.

However, if the risks outweigh the benefits and there is no way to mitigate them, the EPA will propose to cancel the compound's registration, and it will publish a "notice of intent to cancel" in the *Register,* as well as sending copies to the companies who have registered the substance. If the proposal to cancel is appealed by these companies, as it usually is, the process then becomes a long, complex, legal one, involving hearings before an administrative law judge, not a scientist, who considers all the evidence, pro and con, and eventually renders a decision. During this process the compound can still be used.

Suspension of registration is much more immediate. Anytime the EPA obtains reliable information—not necessarily hard scientific evidence—indicating that a pesticide may pose an "imminent hazard" to the environment or human health, it can order the registration of a pesticide or certain uses of it suspended. A brief hearing is held, and if the suspension order is approved, the proscribed use is halted; the whole process takes only about a month. In special cases, if the EPA feels that the danger to human health may be a particularly serious one, it can issue an *emergency* suspension order, in which case the proscribed use(s) stops right then and there, even *before* hearings of any kind are held. This action has only been taken once, in the Alsea case mentioned above. Suspension orders automatically expire after one year; any suspension order, however, also automatically triggers the full cancellation proceedings.

Like all regulatory laws, FIFRA is by no means perfect, and has more or less constantly undergone amendments over the last decade to adapt to the deficiencies that its implementation has revealed. One aspect of the act that has caused a great deal of controversy is its attempt to protect manufacturers' trade secrets while at the same time guaranteeing public access to health and environmental safety data from manufacturers' toxicology tests. Unfortunately, the framers of the act did not foresee that these two kinds of information would tend to be completely intermingled in the documents that EPA receives from manufacturers. Tests on lab animals commonly allude to formulas, inert ingredients, yields, and other kinds of confidential information, which has usually had the result of assuring that the entire document is labeled confidential and made unavailable to the public.

According to the EPA this problem is slowly being resolved. Manufacturers are learning to separate trade secrets from health data, or are using codes to identify these parts of toxicology studies that they do not want to become public knowledge. The EPA is beginning to catalogue data sources for certain compounds, and is slowly publishing standards in which lists of data sources pertaining to the safety of generic compounds are cited. The EPA has also made it clear that in emergency cases, where human health is at stake and a physician may need to have immediate access to toxicology information, the trade secret clause will not be allowed to stand in his or her way.

At the end of 1980, responding to widespread criticism that the EPA had acted hastily and unwisely in the Alsea affair, Congress amended FIFRA in an ominous way. From now on any enforcement actions contemplated by the agency must first be submitted to Congress for review. If Congress doesn't like what it sees, it has sixty days in which to veto the proposed actions. This amendment places yet another stumbling block in the way of an agency that already has significant difficulties acting decisively, and in essence assigns ultimate responsibility for pesticide regulation to Congress, a body that is far less insulated than even the EPA against the pressures of special interest groups, like agribusinesses and chemical companies. As of this writing, Congress has not yet chosen to exercise this veto, but when it does those who feel this amendment is illegal will in all likelihood seek to test it in court.

PESTICIDE EXPORTS

The only pesticides manufactured in the United States that do not need to be registered are those intended solely for export—a sizable portion of our annual production. Though exported pesticides are exempt from FIFRA registration requirements, they have to be labeled in virtually the same way as registered compounds and the label must clearly state that the substance is "Not Registered for Use in the United States." Furthermore, FIFRA also stipulates that foreign purchasers who buy pesticides not registered in the United States must sign a statement acknowledging that they understand this fact. A copy of this statement is then sent by the EPA to their government, to inform it that someone in its country is importing—and presumably using—a pesticide that has not been approved for use in the United States. Also, suspension and cancellation information on registered pesticides is forwarded, via the State Department, to appropriate foreign governments and international agencies.

These provisions have been criticized as inadequate, and exporting all kinds of potentially toxic substances, not only pesticides, has become a controversial issue. Just before he left office, President Carter made a special point of signing an executive order that added a number of further restrictions to the process of exporting a range of different hazardous substances, and called for the annual publication of a warning list of such substances to be distributed abroad. One of President Reagan's first actions upon taking office was to rescind this order.

REGULATION OF PESTICIDE RESIDUES IN FOODS

Though the pesticides we use today do not persist in the environment as does DDT, neither do they instantly vanish after application. Depending on a host of interrelated factors—their exposure to sun, moisture, and oxygen, how they are applied, when they are applied, whether they tend to be absorbed into the plant or remain on its surface, and so forth—a pesticide may leave detectable residues in food crops and other food products for anywhere from a few hours to several months. And a few, very few, of the pesticides in use today even rival some of the CHIs in their persistence in the environment.

Regulation of pesticide residues is shared by the EPA and the FDA. Not all pesticides are used on food crops or food products, of course, but for those that are, the Food, Drug and Cosmetics Act requires that a tolerance—a limit on how much residue can safely be allowed in foods when they are consumed—be set. Since much of the data used in setting tolerances is identical to that used in determining whether or not a pesticide is safe to register, Congress assigned the job of calculating tolerances to the EPA when it was formed. The FDA, which was responsible for the entire job of regulating pesticide residues in foods before 1970, is still in charge of enforcing the tolerances set by the EPA. In some states, like California, the enforcement job may be handled by state food and agriculture departments.

In order to set a tolerance the EPA conducts two kinds of reviews of the data obtained from toxicology tests performed by the pesticide manufacturer. The first kind of review is a chemistry review. It focuses on the compound's chemical identity, its similarity to other substances, what happens to it chemically when it is applied to food crops, and how much of it or its byproducts are left in a food crop or product after certain periods of time have elapsed. Typically investigators find that most of the chemical breaks down fairly rapidly and disappears, but that traces tend to remain for much longer intervals. The amount that would probably be left when the crop was harvested, or the product purchased, called an "efficacy level," is identified. This is usually the residue that a consumer would ingest.

In the second kind of review, toxicologists examine the data to determine the "no-effect level" (NOEL), the dosage below which lab animals show no observable effect from exposure to the compound. Then a safety factor, usually 100, is applied because, as we explained in Chapter 1, humans may be more sensitive to the compound than lab animals and because there are especially susceptible groups in any population—the elderly, the sick, and the very young. The resulting figure is then labeled an "acceptable daily intake," or ADI. Typically ADIs are much larger than efficacy levels—in other words, the amount that toxicologists predict will be safe to consume each day is far more than the residues actually left on or in the foods we consume.

To then set the tolerance for a specific crop or food, say carrots, the EPA begins by checking USDA figures on what percentage of the average diet carrots represent—say 0.48 percent. If the average person eats 1.5 kilograms of food in a day and carrots represent 0.48 percent of that person's diet, then he or she eats 7.2 grams (about one-quarter of an ounce) of carrots a day. Of course, many of us eat more than that and some of us eat less—the figure is only an average. If the efficacy level for carrots for the compound under review is, say, 0.1 ppm (equivalent to 0.1 mg/kg), then the amount of pesticide residue an average 60 kg person will consume in a day is calculated as 0.1 mg/kg (efficacy level) $\times$ 0.0072 kg (average amount of carrots eaten in a day) $=$ 0.72 μg.

If this amount is well below the ADI, as is usually the case, then there's no problem, and the tolerance will be set at or near the efficacy level. If, on the other hand, this amount is close to the ADI, the EPA may require the manufacturer to make changes in the way the compound is made or applied so that a lower residue is left, or the agency may decline to set a tolerance at all, thus effectively blocking the pesticide's use on that crop or in that product. If the pesticide is used on a number of different foods, as many are, tolerances are set for each one in such a way that the combined residues from all of them will not exceed the ADI.

The FDA then spot-checks foods to make sure that the actual residues remain well below the established tolerances. In the vast majority of cases they do. For example, from 1965 to 1974 only 1 percent of all the food crops tested by the FDA contained any trace of 2,4-D, and from 1974 to 1977 none did. From 1964 to 1970 the highest residues of carbaryl—one of the most common pesticide residues found in foods—was equal to only about one-tenth the ADI.[31]

When the FDA does find crops or food products with illegally high residues, it can request that any portion already distributed be recalled. Contrary to popular belief, however, the FDA does not have the power to *order* a recall. It can only inform the company involved that one of its products is contaminated, advise it to recall outstanding amounts, inform the public about the potential hazard, and monitor the progress of the recall. Companies are usually happy to cooperate, if only because they have a healthy fear of large damage settlements in product liability suits. If they resist the recall, the FDA must obtain a court order before it can seize and destroy the crop or product. In some cases it can hold up a crop that it suspects is contaminated while a court order for its seizure is issued, but if the crop is already being distributed it is virtually impossible to retrieve it from truckers, grocery stores, and consumers.

There are a number of safety factors built into the tolerance-setting process. Typically, as we mentioned above, the ADI is far higher than the efficacy level, even after the safety factor of 100 has been applied. Since the pesticides we use today are relatively nonpersistent in the environment, by the time a crop has been harvested, stored, transported, stored again, possibly processed (washed, cleaned, heated, frozen, etc.), stored yet again, transported, stored, purchased, stored, and finally eaten, its residues will have dropped even further. Moreover, since the pesticides we use today are also much less likely to accumulate in animal tissues, the tiny residues that are normally ingested are quickly excreted in most cases.

On the other hand, the process is plainly not perfect. For one thing, the USDA dietary averages that ADIs are based on do not reflect the enormous diversity of food consumption patterns in this country. The USDA may calculate the average consumption of avocadoes is one-half every six months, but some people eat that much every day. Some variations from the average may be so extreme, especially in the case of reducing diets, as to bring a person's residue ingestion close to the ADI or even above it.

For another thing, the ADI system works reasonably well for individual pesticides, but there is no systematic approach for estimating the combined effect of all the various different pesticide residues each one of us may ingest in an average day's diet. In other words, we may be well within the ADI limits for any number of separate pesticide residues, but what about their interaction with one another in the field, factory, or in our bodies? What if pesticide X interacts synergistically with pesticide Y in a way that augments its effect, making what would otherwise be a harmless amount act as though it were the equivalent of ten, fifty, or one hundred times itself? According to the EPA some work is being done in this area, but not enough. It is clearly an area that deserves attention.

There are other problem areas. As we explained above and in Chapter 1, we still don't know enough about the chronic effects of most older pesticides and won't until the EPA has developed standards for them during the next two decades. Yet tolerances have been set for these compounds despite this gap in our understanding of their full potential to cause toxic effects. We are already beginning to learn that some of these tolerances have not been protecting us from the long-term effects of exposure to these substances.

For example, in 1974 the National Cancer Institute warned the EPA that a pesticide known as EDB (ethylene dibromide) appeared to be an animal carcinogen. EDB was first registered in the United States in the 1950s and has been used extensively as an insecticide on stored grains, as a nematocide for a number of crops, as a citrus fumigant, and in a number of other ways. Traces of its byproducts appear in a wide variety of foods, and tolerances for them have existed for some time. Until recent advances in detection technology, however, EDB itself could not be detected in foods and it was assumed that it was not present. We now know this is not the case.

The more the EPA looked at EDB, the more convinced it became that the compound was not just carcinogenic but also a potent mutagen and a cause of a variety of reproductive disorders in rats, mice, and other lab animals. (Ironically, EDB was one of the substitutes that many farmers turned to when DBCP's registration was canceled after it was discovered to be a male sterilant in 1977.) The major public exposure to EDB today is from the wheat used in baking bread. According to EPA calculations, the cancer risk associated with this exposure is somewhere between a high of 6.28 in 10,000 and a low of 2.6 in 1 million.[32]

On December 14, 1977, the EPA published an RPAR notice for EDB in the *Federal Register,* proposing to cancel the use of EDB in stored grains and a number of others, as well as significantly alter the conditions of uses that would be allowed to continue. As of this writing, although the toxicity of EDB and the entire process of how EPA assesses the risks of exposure to toxic substances were embroiled in controversy, the agency was moving ahead with its proposal to cancel certain uses of EDB.

Another problem is that like the EPA and every other regulatory agency, the FDA does not have sufficient enforcement staff to make sure that all foods always comply with established tolerances. It only has the manpower to spot-check crops and food products, and some foods with illegally high residues inevitably slip past it. In fact, the actual amount of food inspected by the FDA is less than 1 percent of everything we eat.

No doubt some foods we eat contain illegal levels of pesticide residues, but most do not, and even those levels we do ingest will nearly always be far too low to harm us. Nevertheless, to be as safe as possible, consumers should make a point of washing and peeling produce. This won't be effective for everything, since many pesticides are applied when the crop is first forming and residues become incorporated into its flesh. For many other kinds of pesticides and crops, especially green leafy vegetables like lettuce and cabbage, washing is surprisingly effective. For example, tests have shown that carbaryl residues on lettuce drop from 30 ppm to less than 1 ppm after a simple wash in warm water.

7

HAZARDOUS WASTES

It's no secret that Americans are wasteful. We throw out enormous amounts of trash each year, much of which we could reuse in ways that would save us money and cut down on pollution. Agriculture, mining, and manufacturing are our biggest waste-makers, generating billions of tons of wastes a year that must be disposed of, somehow. Individuals contribute their share, too—at least half a metric ton of cans, plastic, paper, disposable diapers, and old chicken bones for each and every one of us per year.

Historically, America has been such a land-rich country that we could afford to dispose of our wastes by just dumping them on the nearest piece of unwanted ground. According to the EPA, in 1978 we were using some two hundred thousand hectares for nothing but disposing of the wastes from our cities. But the era of the open dump is rapidly disappearing, for a number of reasons.

First, the traditional dump was a health hazard. It attracted disease-carrying flies, rodents, and other scavengers, and the decomposing organic matter in garbage bred potentially hazardous bacteria, as well as generating toxic gases like hydrogen sulfide and methane. In response to these problems we slowly developed "sanitary landfills"—basically the same as dumps, but compacted and regularly covered with a layer of dirt to prevent the spread of disease and odors. Sanitary landfills are still common in this country, and will be for some time to come, but their future is also limited. Suitable sites are growing harder and harder to find, the energy costs of transporting garbage to them and maintaining them is soaring, and, most important, we can no longer afford to waste the reusable resources—glass, paper, aluminum, minerals, combustible gases, and hundreds of other substances and commodities—that we traditionally buried in them without a second thought. Our slow realization that it costs less and pollutes less to recycle rather than always start over from scratch with new resources has only confirmed the demise of the classic landfill dump. The future of waste disposal is clearly one in which every resource and commodity that can be recovered from our wastes and recycled will be and much of whatever is left will be burned in huge RDF (refuse-derived fuel) power plants that will provide the country with millions of kilowatts of electricity, all generated from garbage. The EPA calculates that the potential energy savings from RDF plants is equal to some 400,000 bbl of oil per day, enough to light every American home and office.[1]

Concern about America's waste problem has been growing ever since the mid-1960s, when the first national legislation aimed at coping with the problem of ever-increasing amounts of waste—the Solid Waste Disposal Act of 1965—was passed. A decade later, in 1976, a more comprehensive waste disposal law—the Resource Conservation and Recovery Act—was enacted by Congress. Though it contained the basic legal framework for our current system of regulating toxic wastes, its major thrust, as the name implies, was to encourage the recovery and reuse of the millions of tons of valuable resources we discard, as well as to stimulate a search for more efficient and sanitary methods of disposing of wastes than the traditional open dump.

Until quite recently, however, this concern focused almost exclusively on problems of volume and site location, not toxicity. Then, in August 1978, an environmental health disaster that had been more than three decades in the making hit the headlines, jolting us into a sudden awareness of an especially ominous kind of disposal problem—hazardous industrial wastes, particularly toxic chemicals.

On August 2, 1978, under the headline UPSTATE WASTE SITE MAY ENDANGER LIVES, *New York Times* reporter Daniel G. McNeil, Jr., broke the story of what was to become the most sensational environmental issue of the decade: Love Canal. The very next day New York State Health Commissioner Dr. Robert P. Whalen, citing "growing evidence of subacute and chronic health hazards as well as spontaneous abortions and congenital malformations," urged the evacuation of some twenty to forty families with pregnant women or young children, all living on the periphery of the old dump site. In 1978 and again two years later, on May 21, 1980, President Carter declared Love Canal a national disaster, marking the first time in the nation's

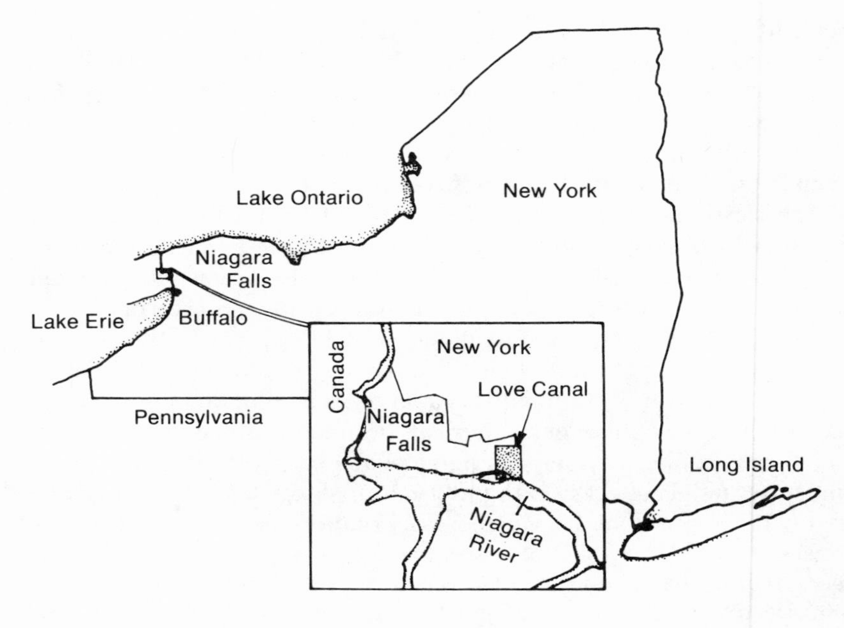

Figure 22

history that anything other than a natural catastrophe had earned this dubious distinction. In the last few years Love Canal has come to symbolize our belated recognition that the same postwar boom in the use of synthetic chemicals that has done so much to improve our lives has also left us with a toxic and expensive legacy. The EPA estimates that there are some twelve hundred to two thousand other potential Love Canals scattered around the country—chemical time bombs waiting to go off, as many observers have called them. The cost of cleaning them up will be at least $20 billion and may run twice as high as that.

Like so many other environmental crises in industrialized countries, Love Canal had its genesis in an entrepreneurial vision.* In the late nineteenth century businessman William T. Love conceived of a bold, imaginative plan to dig a canal linking upper and lower Niagara Falls, thus creating a seven-mile millrace that would provide plentiful, cheap hydroelectric power to the area. But while Love was dreaming, technology was marching on. The discovery that alternating current made the transmission of electricity over long distances

*The best firsthand account of the Love Canal story that we know of is in Michael Brown's *Laying Waste: The Poisoning of America by Toxic Chemicals*, Washington Square Press, New York, 1979.

by wire feasible doomed his project before it ever really got started. Abandoned by Love, the canal trench, some 900 meters in length, 18 meters in width, and between 3.5 and 12 meters in depth, remained nothing more than a long, narrow hole in the ground until the early 1940s.

In 1942 the Hooker Electrochemical Corporation (forerunner of today's Hooker Chemicals and Plastics Corporation) leased the trench site from the city of Niagara Falls and began using it as a dump for chemical wastes.* By 1953, when it stopped using the dump, Hooker had poured nearly twenty thousand metric tons of these wastes into William Love's legacy.

At this time, there were few houses around the old trench, and in many ways the site was a good one for this kind of dump. Although the top few meters of soil in the area were sand and silt, from 1.5 to some 7 meters below the surface the soil was thick clay, an ideal material for containing wastes. William Love's trench excavation had in essence scooped out a giant bathtub in this clay, from which leaching of wastes was highly unlikely. Below the clay was a layer of "glacial till," also highly resistant to the migration of wastes. Both the American Institute of Chemical Engineers and the EPA are on record as stating that Hooker's choice of the Love Canal site was, geologically speaking, consistent with standards of the time, and would in many ways be acceptable today. A clay-lined trench is still regarded as a perfectly acceptable way to dispose of numerous kinds of wastes, including many that are quite toxic, but current practices and laws would require a company like Hooker to install a drainage system around the dump to insure that if wastes did begin to migrate away from the site they would be collected and diverted into some kind of backup repository.

Also, Hooker's practice of simply pushing waste-filled drums off the back of a truck into the trench, where they fell in a haphazard way, wouldn't be permitted now. As drums buried in a random way age and corrode, they begin to leak and flatten out under the pressure of the earth and other wastes overhead. This leads to settling of the soil, which in turn can cause large fissures through which waste liquids and fumes may escape. Today there would also be elaborate groundwater monitoring requirements to make certain that local aquifers were in no danger of being polluted by escaping wastes. And finally, of course, no landfill holding toxic chemicals would be

*In the mid-1970s Hooker was purchased by Occidental Petroleum, which also owns the chemical plant discussed in Chapter 3 in connection with the DBCP tragedy.

allowed anywhere near an area as heavily populated as the Love Canal neighborhood became in the years after Hooker stopped using the dump.

By 1952 Hooker had filled the north and south ends of the trench with chemical wastes, and the city of Niagara Falls had filled the middle with municipal garbage. Dumping halted, and the trench was filled in. Whether its cover was a thick clay cap, as Hooker claims, or just dirt from the edges of the original excavation, backfilled in by city workers, is now a matter of significant dispute. A clay cap would have sealed the wastes in a virtually impenetrable clay tomb, preventing their escape in any direction. Test borings taken in recent years show no evidence of clay in the top couple of meters of fill at the site.

Though the land around the trench was only sparsely populated when Hooker first began dumping there, after the Second World War the city of Niagara Falls began to grow. The postwar baby boom hit, and the Board of Education began eyeing the old dump site as a good place to build a new grade school. In 1952 they notified Hooker that they were interested in the sixteen-acre Love Canal property. Hooker pointed out that the site had been used for chemical waste disposal, a fact that might render it unsuitable for development. The board persisted, however, and Hooker soon acquiesced, though not without stipulating that the deed acknowledge the fact that the site had been used as a chemical dump and contain a provision releasing Hooker from any liability for injury caused by the wastes buried there. Apparently, Hooker had significant misgivings about the board's plans from the start.

Five years later the board had built its school and was moving to sell parts of the Love Canal site to housing developers. Hooker again pointed out that the kind of excavation required for streets, sewers, basements, and other residential construction might disrupt the clay barrier around the wastes and lead to serious exposure hazards. These warnings may have delayed development of the area for a few years, but by the early 1960s the land around the old canal was almost completely filled in with new homes.

During the 1960s and early 1970s it was no secret to most of the twenty-five-hundred-odd people living in the Love Canal neighborhood that the trench had once been used as a chemical dump, but they paid little attention to the occasional chemical fumes in the air or the iridescent sheen in the seepage in storm drains and old streambeds in the area. After all, Niagara Falls is a chemical town, and Hooker is one of its largest employers.

But in 1976, after many years of unusually heavy snows and rains, the old trench apparently filled with water and began to overflow its clay banks, something that probably wouldn't have happened if a clay cap had in fact been placed on top of it. It also seems likely that seepage did occur, as Hooker had warned, along the paths of the sewer and water lines that the new development had required, which had indeed disturbed the clay walls of the old dump. Chemical toxins began to seep out from their crypt. Basement sumps in the area began to fill up with a black, fragrant ooze. Plants and pets began to sicken and die, children came down with mysterious illnesses, and the incidence of miscarriages, stillbirths, and birth defects began to climb.

Prompted by increasing complaints from local residents, health authorities eventually began conducting extensive analyses of the residues leaching out of the old canal and had soon identified some eighty different chemical compounds, including a number of hazardous pesticides, TCE (described in Chapter 5), and even traces of TCDD (described in the previous chapter). About a dozen of the eighty-odd compounds were thought to be carcinogenic in humans; one, benzene, is a known human carcinogen. Further testing of the air, water, and soil confirmed the presence of high levels of chemical toxins in the Love Canal environment. Most alarming of all, an extensive review of birth records revealed that the miscarriage rate in the area was at least 50 percent higher than normal.

Armed with this information, Dr. Whalen made his dramatic announcement in August 1978. A few days later Governor Carey authorized an extension of the evacuation to include another two hundred families in the immediate vicinity of the canal. The school was closed and a chain-link fence was erected around the six most contaminated blocks. President Carter's declarations made federal funds available to help cover the costs of medical and environmental sampling, cleanup costs, and evacuation expenses for about one thousand families.

Since 1978 evidence of another four hundred chemical residues have been found in the Love Canal environment. The EPA is carefully monitoring the groundwater in the area, which so far appears to be uncontaminated. State and federal outlays for assistance to local residents and for testing, cleanup, and monitoring costs have mounted to over $75 million. A clay cap and an effective drainage system have finally been installed in the old trench. Seepage from the canal is now processed through a sophisticated purification plant.

However, despite overwhelming circumstantial evidence, no one has yet conclusively connected the undisputed presence of high levels of toxic chemicals in the Love Canal area to any of the abnormal health problems experienced by Love Canal residents. The relationship seems obvious, but has yet to be demonstrated scientifically.

Lawsuits against Hooker by state and federal authorities now amount to over $780 million. Additional private claims against Hooker, the city of Niagara Falls, Niagara County, the Board of Education, and others now amount to well over $15 billion.

THE SCOPE OF THE PROBLEM

Of the billions of tons of wastes we generate each year from all sources, only a relatively small proportion—some 57 million metric tons, according to the EPA—is toxic. Nearly all of that amount comes from industry, well over half from chemical companies.

Strangely enough, although the technology for safe disposal of most hazardous wastes is well-developed, the EPA also calculates that no more than every tenth ton of hazardous waste is disposed of safely. In effect, what this means is that we are annually allowing over 50 million metric tons of potentially deadly waste compounds to contaminate our environment and raise the specter of serious harm to our collective health.

That amount is rising steadily each year, partly because our industrial production is rising, but also because environmental protection laws like the Clean Air Act and Clean Water Act have forced industries to collect much of the waste emissions and residues that they were once able to release freely into the biosphere. And once collected, that waste has to be stored, treated, and eventually disposed of, somehow. Until recently by far the most popular method of getting rid of it, despite the existence of perfectly sound alternatives, was to dump it in unsafe and ineffective holes in the ground.

Why this obtuse reluctance to make use of better disposal techniques? In a word, *money.* It has always cost far less for industries simply to pour their wastes into dumps, ponds, pits, and so forth and leave it there than to incinerate it, distill it, precipitate it, oxidize it, adsorb it with special carbon filters, feed it to microbes, or treat it with any of the many other methods available for rendering it harmless.* For example, the EPA figures that simply collecting wastes in

*As we pointed out at the end of Chapter 3, PCB disposal has recently become something less of an issue since the EPA has granted operating permits to two incinerator companies and one chemical reagent system developed by the Sunohio Com-

a "surface impoundment" costs no more than $14 to $180 per metric ton. By comparison, incineration can cost anywhere from $75 to $2,000 per metric ton.[2] Incinerating a tank car of toxic wastes can cost as much as $10,000. In the period from 1975 to 1978 the EPA found that only 6 percent of all the hazardous waste generated in this country was incinerated; 78 percent was disposed of in unsafe landfills and surface impoundments.* So, it has traditionally been cheaper to dispose of hazardous wastes unsafely than safely, a situation that has changed only in the last year.

Industry representatives sometimes respond to this argument by pointing out that most hazardous wastes, regardless of the *method* of their disposal, are disposed of on the property of the firm that generates them. Technically, this is true—most hazardous industrial wastes do not leave the property of the company that produces them. But this response overlooks the interdependence of the biosphere. Wastes poured into an underground shaft may eventually leach out into a nearby aquifer, toxic liquids may overflow the sides of their ponds and run off into nearby surface waters whenever it rains, and a waste dump that explodes into flames can spread toxic smoke and fumes over wide distances, far beyond the confines of the owner's legal property lines.

None of these possibilities is at all hypothetical. All have actually occurred at one or more places within the United States over the last few decades. For example, from 1964 until very recently a subsidiary of a major aeronautics firm in Sacramento, California, involved in the manufacture of paint constituents, herbicides, and pharmaceuticals, dumped at least two dozen different kinds of organic and inorganic contaminants, many quite toxic, into waste lagoons, deep shafts, and septic tank systems that were not adequately shielded from local groundwater supplies. In January 1979 concentrations of TCE and dozens of other pollutants at levels as high as 8,360 μg/liter were discovered in test wells, and some nearby drinking water wells have since been closed to human consumption. Another example: A 4-million-liter pool of toxic wastes in Epping, New Hampshire—so toxic that anyone working near it must wear special breathing gear—threatens to overflow its banks every time the area receives a heavy rain. In April 1980, a chemical waste dump in Elizabeth, New Jersey,

---

pany. But all of these processes are significantly more expensive than the favored PCB disposal methods of the past—simply dumping it out on the ground, burying it in old drums, or letting it flow into local rivers and streams.

* A landfill is basically a hole in the ground; a surface impoundment has some other structural support, such as concrete, besides the soil to help contain wastes.

caught fire and blew up with a series of explosions audible in downtown Manhattan. Three months later fires broke out at similar dump sites in Carlstadt and Perth Amboy, New Jersey.

In defense of its industry the Chemical Manufacturers Association (CMA), representing the top two hundred American chemical firms, points out that as early as 1971 the industry had established CHEMTREC (Chemical Transportation Emergency Response Center), an emergency information service developed to respond to reports of chemical spills and other emergency situations. In 1979 the industry set up a similar service, called the Hazardous Waste Response Center, specifically to help in the fight against "orphan" hazardous waste sites. The CMA also likes to note that the chemical industry's capital investment in pollution control of all kinds now exceeds $6.7 billion, more than that of any other industry in the country. By the mid-1980s waste disposal alone will have cost the industry another $10 billion, according to the CMA.

And in many ways it should be acknowledged that the chemical industry has become much more sensitive to the problems of hazardous wastes recently. With the exception of a few flagrant violators, it has not been the major chemical companies who have been responsible for the worst cases of hazardous or irresponsible dumping. Instead, it is far more often the smaller companies and the middlemen—manufacturers who purchase chemicals for use in their operation and then hand the wastes on to someone else, all too often a fly-by-night disposal company with no scruples about dumping wastes illegally and unsafely—who cause the worst problems. Large corporations like 3-M, Monsanto, Dow, and Union Carbide have a reasonably good record when it comes to retrieving drums bearing their labels from orphan dump sites and redisposing of them safely somewhere else. Moreover, these same firms have pioneered many of the more advanced methods of disposing of hazardous wastes that we are beginning to make greater and greater use of, and are actively cooperating in the organization of waste exchanges where one company's wastes become another company's raw material.

However, despite these positive steps on the part of some of the top firms in the chemical industry, the bottom line is that American industry as a whole still spends very little—a small fraction of its gross revenues—on handling the problem of the hazardous wastes it generates. According to EPA estimates, the industries primarily responsible for the generation of hazardous wastes earn annual revenues of some $154 billion.[3] Chemical industry sales account for nearly all of this total, some $148 billion. In 1978 these industries spent a collective $155 million on waste disposal, or about 0.1 cent for every dollar they took in. Regulations that went into effect in 1980 will push that amount up only very slightly, to about 0.5 cents per year per dollar of gross revenues.

THE RESOURCE CONSERVATION AND RECOVERY ACT

In the fall of 1980 Americans were treated to the grim spectacle of what appeared to be an epidemic of illegal dumping of hazardous wastes all over the country, especially in the heavily industrialized areas of the Northeast. Unscrupulous waste disposal contractors, some reputedly linked to organized crime, were making a quick buck by taking toxic wastes off the hands of panicked businessmen and dumping them anywhere they could get away with it—in open fields, alongside highways, on public roads, in shopping center parking lots, in warehouses leased under false names and then abandoned when they were full, in railroad cars loaded with waste-filled drums then shipped COD across the country to fictitious addresses elsewhere. To the millions of viewers across the nation, watching reports on the evening news, it seemed as though someone had taken a large stick, thrust it into an anthill, and stirred.

That someone was Congress and the EPA, and the stick was a new set of regulations making those who generate hazardous wastes responsible for disposing of them properly, no matter whose hands they may have passed into after they were generated. The new rules are part of comprehensive solid* waste disposal legislation, the Resource Conservation and Recovery Act, known informally as RCRA (pronounced "reck-rah"). First suggested by the EPA in 1973 and enacted by Congress in 1976, RCRA's provisions for hazardous waste disposal were not officially implemented until November 19, 1980. The rash of illegal dumpings that year was the result of a widespread last-minute rush to beat the deadline on the part of environmentally insensitive businesses across the country.

The Resource Conservation and Recovery Act (Public Law 94-580) is a set of comprehensive regulations covering the generation,

*For reasons that are a mystery to us, the term "solid" under RCRA is defined so broadly as to seem almost meaningless—it officially includes not only wastes that are solid but also those that are "liquids, semisolid, or contained gaseous material." Anything disposed of on or in the land is considered a "solid" waste. About the only thing RCRA specifically excludes is nuclear waste and domestic sewage or industrial point source pollution regulated under the Clean Water Act.

storage, and disposal of solid wastes of all kinds, including both hazardous and nonhazardous substances. RCRA was enacted to complement previous legislation protecting air and water quality, thus broadening the EPA's power to protect the entire environment. It is also intended to promote conservation and recycling.

Subtitle C of RCRA defines "hazardous" as anything that is flammable, explosive, corrosive, or toxic. The EPA also regularly updates a list of some two hundred industrial processes known to produce toxic wastes. Anyone who generates more than one thousand kilograms a month of any kind of waste at all is obliged by RCRA to consult these lists or conduct certain tests to determine if the wastes are hazardous; RCRA's basic presumption is that they are, until proven otherwise. Moreover, a waste generator cannot just check the lists and perform the tests once—the law stipulates periodic retesting since the chemical composition of wastes can change over time owing to spontaneous reactions or changes in manufacturing processes.

FROM CRADLE TO GRAVE

RCRA also mandates a tracking system of manifests and reports to the EPA that must be filled out by generators, transporters, and by those who ultimately store and dispose of hazardous wastes. It was this "cradle-to-grave" manifest system that caused all the hysteria and eleventh-hour dumping in 1980 as businesses rushed to get rid of wastes before they came under this comprehensive and expensive new system.

Under RCRA, from the day a drum of toxic wastes is produced to the day it is disposed of it must be accompanied by its manifest, and reports from those who generated it, transported it, and stored, treated, and disposed of it must be submitted to the EPA. Using code numbers for each generator, transporter, and storage facility, the cradle-to-grave manifest system is designed to interconnect these three major links in the waste disposal process to make sure that wastes can never be surreptitiously "orphaned," and that if they are, their source, carrier, and destination can be traced. Manifests also must state the shipping description, hazard class, quantity of waste, type of container, and special contacts and handling information in case of emergencies like accidental spills.

RCRA stipulates that anyone who stores hazardous wastes for more than ninety days, even on the same site where they were generated, automatically becomes a "treatment, storage, and disposal facility," or TSDF for short. RCRA requires TSDFs to have permits, and imposes extremely demanding standards on them prescribing their location, design, operating procedures, emergency procedures, personnel training, monitoring and control systems to prevent air and water pollution, and security systems.

The financial requirements of TSDFs, prescribed by RCRA, are stringent. TSDF operators must arrange for $10 million in liability insurance, must post a cash deposit to cover the costs of their eventual closure, and must make annual deposits into a "postclosure care" trust fund, used to monitor and maintain them in a safe condition for thirty years after they shut down. As of this writing, however, the Reagan administration was moving to weaken these requirements significantly.

RCRA also contains a clause that authorizes the EPA to seek a court order to shut down immediately any handling of hazardous wastes that is posing an "imminent and substantial endangerment to health or to the environment." Violators of RCRA regulations face stiff penalties: up to twenty-five thousand dollars per day in fines and a possible one-year jail sentence for first offenders; up to fifty thousand dollars per day and two years in jail for subsequent offenses. Some states have even tougher toxic waste laws. New Jersey, for example, can fine violators up to fifty thousand dollars per day and can jail them for up to *ten* years.

THE SUPERFUND

As comprehensive and foresighted as RCRA appears to be in many ways, its creators overlooked one major problem. The power of the EPA to do something about a toxic dump that posed an imminent hazard was limited to bringing suit against its owner to clean it up. If there is no owner, or the owner is financially unable to cover the costs of correcting the hazard, then the costs fall on the shoulders of the taxpayer. In all too many cases the waste dumps that do pose toxic hazards are "orphans"—their owners have disappeared, cannot be traced, or are not in a position to pay for the cleanup if they are identified and located. For this reason it became apparent to the EPA and Congress in the late 1970s, as more and more orphan dumps were uncovered, that a special fund would have to be established to cover these costs, and that most of its revenues should come from the industry that produced the most hazardous wastes and gained the most profit from them—the chemical industry.

However, despite its professed interest in managing toxic waste disposal safely, the chemical industry had no wish to see cleanup legislation passed that would saddle it with additional burdensome federal regulations as well as cut into profits. Under the CMA's orchestration it set about to defeat or at least weaken the impending cleanup legislation.

To do this, a two-pronged campaign was launched. On the one hand, phone calls and letters from chemical company shareholders and officers alike poured into congressional offices from all around the country, urging members of Congress not to "overreact" to public hysteria by approving unjustly harsh legislation. Second, a program of carefully placed contributions from chemical company political action committees to the campaign funds of members of Congress running for reelection was initiated. According to Ralph Nader's Congress Watch, $2 million was donated to thirty-two senators and fifty-five representatives, many holding key positions on committees that oversee environmental regulations. Contributions ranged from five thousand dollars to almost thirty thousand dollars.

By late fall 1980, adjournment of the Ninety-sixth Congress was nearing and two separate versions of a cleanup bill had been passed, one by the House and one by the Senate. Both were now ready for consolidation by a joint House-Senate conference committee. The House bill, the milder of the two, called for a cleanup fund of $1.2 billion, 75 percent of which would come from fees charged on the raw feedstock substances used by chemical companies. The remaining 25 percent would come from general tax revenues. Only cleanup costs would be covered. No provisions for reimbursement to victims whose health or property was damaged by waste dumps was included.

The chemical industry, though still fundamentally opposed to any cleanup bill at all, especially one that implicitly assigned the liability for the costs of cleaning up old dumps to it, went on record then as favoring an early version of the House bill, one drafted by the Committee on Interstate and Foreign Commerce, which called for a fund of only $600 million, half paid by industry fees and half by the taxpayer. The CMA made it clear that it was unequivocally opposed to the much tougher Senate bill.

The Senate superfund cleanup bill, highly favored by environmentalists, the Carter administration, and state and local governments alike, originally contemplated a $4.1 billion fund, 88 percent of which was to come from industry sources and 12 percent from the Treasury. A full one-third of this fund was to be devoted to compensating waste dump victims for health and property damage.

However, by the time it came to a vote in the Senate, the CMA lobbying campaign had paid off handsomely. The bill was slashed to $1.6 billion, the industry share dropped to 85 percent, and provisions for victim compensation were dropped. Only the costs of cleanup and property damage were retained. (That $1.6 billion total is only about 8 percent of the amount that the EPA calculates will be required to clean up the remaining twelve hundred to two thousand hazardous waste sites dotted across the country.)

On December 11, 1980, the final bill, formally known as the Comprehensive Environmental Response, Compensation, and Liability Act of 1980, was signed into law by President Carter. It establishes the $1.6 billion fund for a period of five years, to be administered by the EPA. Of the total amount, $1.38 billion will come from fees levied on basic materials widely used in the chemical industry. Funds will cover the costs of cleaning up waste dumps and spills, including those on water. Where dumping can be traced back to a specific firm, the law authorizes the EPA to sue to recover cleanup costs. The law also contains provisions enabling the President to declare an environmental emergency in waste dump spills and Love Canal-type disasters and order evacuation from areas where wastes pose imminent threats to the health of local residents.

The biggest obstacle to developing an effective nationwide system of safe toxic waste disposal sites seems to be sociological rather than technical. Everyone agrees that we need these facilities, but no one wants them built next door. And this resistance, as reasonable as it may be in many ways, has had the harmful effect in some places of prolonging the use of inadequate landfill disposal methods. In California, for example, the EPA has had to extend its deadline for closing the only landfill site permitted to dispose of PCB equipment because of local opposition to proposed PCB incinerator sites. One possible solution to this dilemma may be incineration at sea.

OCEAN INCINERATION

Since 1969 two West German companies have been successfully using two ships converted to burn liquids at high temperatures to dispose successfully of toxic wastes. The older of the two ships, the *M/T Vulcanus,* is operated by a Dutch company, Ocean Combustion Services (OCS), a subsidiary of the well-known German Hansa Line. The other, the *Mathias II,* is a thirty-five-hundred-ton tanker owned and operated by Industrie Anlage, a West Berlin firm.

The *Vulcanus* is a medium-sized (102 meters long) former freighter altered to carry special tanks and two large incinerators at the stern, both constructed for temperatures of 1,250°C or better. In 1974 and 1977 the *Vulcanus* successfully burned chlorinated hydrocarbon wastes under EPA supervision in the Gulf of Mexico. Then, again, on July 14, August 6, and August 24, the *Vulcanus* burned a total of 10,400 metric tons of Agent Orange at a site in the Pacific Ocean approximately one thousand kilometers west of Hawaii. Each burn took about ten days. Constant monitoring of the emissions from the ship's stacks showed that 99.9 percent of the toxic compounds in the wastes were destroyed, including the as much as 47 ppm TCDD with which they were contaminated.[4] The only emissions of any significance were oxygen, carbon monoxide, carbon dioxide, hydrocarbons, water vapor, nitrogen, and a small amount of hydrogen chloride. Hydrogen chloride can degenerate into hydrochloric acid, but EPA scientists believe the presence of abundant, alkaline seawater near ocean incineration sites will be more than enough of a safeguard to offset this potential hazard.

The absence of hazardous atmospheric emissions from these incinerator ships means that no costly air pollution control equipment is needed in their operation, and this in turn means that ocean incineration can be performed much more inexpensively than land-based incineration—for less than one hundred dollars per ton, far below the one thousand dollars-per-ton median associated with land-based incineration.

However, there are two possible drawbacks to ocean incineration. First, since getting the wastes to the port where an incinerator ship can dock will often involve transportation over long distances, the risk of spills en route is increased. Second, once the wastes are on board the hazards of a spill are in some ways even more disturbing to contemplate. The problems we've witnessed in the past with devastating oil spills at sea may be only a pale hint of the catastrophic damage to marine life that one ocean spill of millions of liters of a virulent toxin could cause.

So far, though, the advantages appear to outweigh the disadvantages. OCS has stated that it plans to station an incinerator ship in the Gulf Coast to serve the American market, and there are signs that a U.S. firm will soon commission its own incinerator ship. Some consideration is apparently also being given to the use of abandoned offshore oil rig platforms for at-sea incineration.

8
RADIATION

It seems fitting to end with a discussion of one technological achievement—radiation—that, perhaps more than any other, epitomizes the environmental health issues this book has focused on: Like so many of the other hazards we've discussed in these pages, radiation is basically a natural phenomenon that industrial cultures have learned to exploit for practical ends. It poses both acute and chronic health hazards, including cancer, mutation, birth defects, and reproductive disorders. Many radioactive substances are extremely persistent in the environment and cumulative in the human body. Much of what we know about their potential to cause illness has been learned from occupational settings. Radiation can contaminate the air, both surface water and groundwater, and the food we eat. The wastes of radiation technology, among the most hazardous our culture has ever generated, create disposal problems that rival or exceed those of any other industry.

Mixed with these drawbacks are the blessings that inevitably go hand in hand with environmental hazards. Among the innumerable beneficial applications of radiation are uses in pest eradication, smoke detection and fire prevention, industrial measurement, soil analysis, electrical power generation, and an ever-expanding array of health care uses, ranging from detecting cavities to treating cancer and powering pacemakers. With radiation the fundamental health tradeoffs of industrial culture—its power to aid and support life as well as maim and destroy it—are as sharply etched as anywhere else.

RADIATION BASICS

Radioactive substances are those whose atoms spontaneously disintegrate, emitting certain kinds of electromagnetic waves or particles in the process. There are four principal types of radioactive emissions: Alpha and beta rays are both particulate emissions. Gamma rays, which, having no mass, are pure energy. (X rays are a type of artificial gamma ray, produced by man-made devices.) The fourth type of radioactive emission, neutron rays, is produced in atomic bombs and nuclear reactors.

None of these types of radioactive emissions can be detected by the unaided senses. They are all harmful because they can "ionize" living tissue, that is, alter its fundamental atomic structure. Most atoms have the same number of positively charged protons and negatively charged electrons, and are therefore electrically neutral. But when subjected to bombardment by ionizing radiation, atoms can lose or add electrons, acquiring a positive or negative charge, and give off or take on the energy released by these exchanges. Although there are compounds in nature, especially certain chemical salts, whose elemental constituents routinely take on positive and negative charges as part of normal bonding processes, for reasons that are not yet fully understood ionization of atoms that are usually neutral can damage them in a number of ways, usually divided into two major subtypes: somatic and genetic. Somatic health effects are those that occur within the life of the irradiated organism. Cancer is the most obvious example. Genetic health effects, resulting from alterations in the DNA of reproductive cells, are those that lead to mutations in future generations.

The principal types of ionizing radiation have widely varying powers of penetration and ionization. Alpha radiation is highly ionizing but lacks penetrating power—a sheet or two of ordinary paper will stop it completely. Beta radiation is somewhat less ionizing, but somewhat more penetrating as well. Gamma and X rays are still less ionizing but have extremely high penetrating powers. The major hazard of radiation from outside the body is due to gamma, X, or neutron rays. Alpha and beta rays are generally harmful only if swallowed, inhaled, or absorbed through direct skin contact.

MEASURING RADIATION

Because radioactivity is essentially a process of natural decay, it tends to lessen gradually over time rather than come to an abrupt

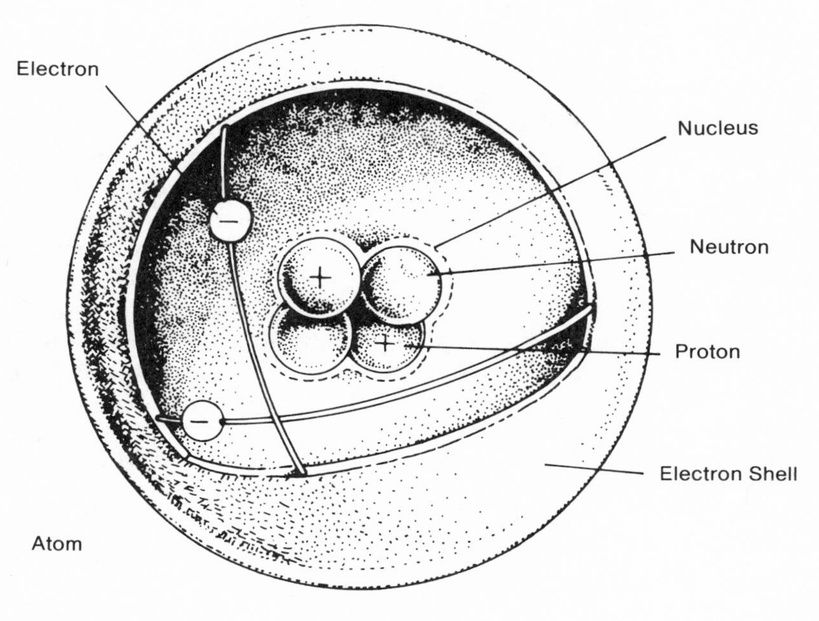

Figure 23

halt, and thus it does not readily lend itself to systems intended to measure its duration from start to finish. For this reason scientists prefer to speak of the "half-life" of a radioactive substance—the length of time it takes for its level of radioactivity to diminish by half, which is always a constant number for that substance and its decay byproducts. There are hundreds of different radioactive elements, and thousands of their byproducts, with half-lives ranging from a few milliseconds to billions of years.

When radioactive substances are absorbed into the body, normal processes of excretion tend to expel them again. Though a small portion of these substances may remain in the body, much is eliminated through exhalation, perspiration, urination, and defecation. The result of the combined action of natural radioactive decay of a substance and its biological elimination from the body is known as the *effective* half-life.

There are a number of different but closely related ways to measure ionizing radiation. The basic unit, the roentgen (pronounced "renkin," after the discoverer of radiation, Wilhelm Roentgen), is a measurement of ionization in air. Specifically, it is the level of X-ray energy that will create 2 billion ions in a cubic centimeter of air. To measure the ionization in other substances, including living tissue, two other units are used: the rad (*R*oentgen *A*bsorbed *D*ose) and the rem (*R*oentgen *E*quivalent *M*an). The rad, which is equal to 100 ergs* absorbed per gram, was originally devised to measure X-ray exposures and is still used for that purpose today. Because X rays and gamma rays are virtually identical, rads can be used to measure them both. But since other kinds of radiation, such as alpha rays, may cause biological effects identical to those caused by X rays at very different energy levels, another radiation measurement was needed that could account for these differences, and thus the rem was developed. For X rays and gamma rays, rads and rems are virtually identical, but for alpha, beta, and other kinds of radiation they are not, and in these cases different numerical factors are used in the calculation of rems, or millirems (thousandths of a rem), abbreviated mR.

ACUTE EFFECTS

Studies of victims of industrial radiation accidents have shown that sudden, intense irradiation of the whole body by gamma or neutron rays causes a distinct pattern of illness known as acute radiation syndrome (ARS), or just plain "radiation sickness." The severity of the syndrome is closely related to overall exposure levels, highly variable individual sensitivity to radiation exposure, age (children are more susceptible to radiation), and overall state of health. Below 50 to 100 rems (100,000 mR), few adults in good health will experience any immediate, overt symptoms.

However, since radiation exposure is cumulative, and many experts argue that it has no threshold level below which exposures are harmless, any exposure at all may add to an individual's risk of developing chronic effects—cancer, mutation, reproductive disorders, birth defects, and a somewhat mysterious shortening of the lifespan.

From about 200 to 400 rems, ARS initially manifests itself as a flulike condition characterized by chills, headache, fever, sore throat, and lassitude. As it progresses, these symptoms are succeeded by weight loss, hair loss, and bleeding gums. Victims usually recover, but slowly: After an illness lasting two months or so, full recovery may take another six months. Though fatalities from ARS have been known to occur at dosage levels as low as 200 rems, the greatest dan-

*A force that has moved a mass of one gram through a distance of one centimeter at an acceleration of one second per second has performed an amount of work equal to one erg.

ger at this stage is probably infection of the respiratory tract, leading to conventional complications like pneumonia.

Up to about 500 to 600 rems the major effect of ARS is on the organs in the body that make blood—the bone marrow, spleen, and lymph nodes—resulting in a decrease in white and red blood cells and platelets. As the dosages rise beyond that level, other, less sensitive systems of the body begin to be affected, first the gastrointestinal tract and ultimately the central nervous system. At a dosage level of 450 rems about half of an exposed human population will die from ARS effects in the absence of appropriate medical care. The appropriate medical response in these cases is supportive—i.e., keeping the patient alive while the syndrome runs its course. Usually this means treating infections and keeping the victims from dehydrating, through the use of intravenous fluids if necessary. If such treatment is available, the human LD_{50} rises to about 600 rems. Exposure of the gonads to 300 rems can cause temporary sterility in men; at doses of 600 rems or more, this sterility becomes permanent.

From 600 to 1,000 rems the gastrointestinal tract becomes the major target of ARS. Although it may also include many of the signs of lower doses, at these levels the primary ARS manifestation is in the form of ulcers, diarrhea, severe abdominal pain, nausea, vomiting, and internal hemorrhaging. Within two to four weeks after exposure, victims of those dosage levels will become lethargic, disoriented, and comatose. At doses above 700 rems few if any exposure victims will survive.

Finally, at extremely high acute doses (e.g., 1,000 rems and up), ARS tends to bypass its earlier targets in the body and attack the central nervous system directly. After a few hours—a day or two, at most—of nausea, vomiting, tremors, convulsions, incoherence, abdominal pain, heavy sweating, hyperventilation, and falling blood pressure, the victim will fall into a coma and die.

Only acute exposures of the whole body to these levels causes ARS. Certain parts of the body, such as the hands, can tolerate significantly higher exposures with only localized damage. Furthermore, some kinds of cancer treatment, such as for bone cancer or Hodgkin's disease, may involve very high levels of radiation, as high as 4,000 to 5,000 rads. But since this treatment focuses on a very restricted part of the body and shielding is used to protect the rest of the body from exposure, the effects are generally beneficial: Cancerous tissue is destroyed without causing excessive damage to adjacent healthy tissue.

Some parts of the body are much more sensitive to radiation than others. The most highly sensitive organs are the eyes, gonads, the thyroid gland, and the bone marrow, spleen, and lymph nodes. Among the least radiosensitive parts of the body are the nerves and muscles.

At the cellular level the fact that rapidly growing, dividing cells are among the most sensitive to radioactivity accounts for the fact that radiation is much more hazardous to children and to the human fetus than to mature adults. It also explains why radiation can be used to treat cancer cells—typified by uncontrolled growth—without doing equal harm to nearby organs or tissues.

CHRONIC EFFECTS

Other studies have taught us much about the long-term health effects that both acute exposures and chronic, low-level exposures to radiation can cause. For example, some five to six years after the explosions at Hiroshima and Nagasaki, the incidence of leukemia among survivors rose to twenty-five to thirty times its level among comparable, nonexposed Japanese populations, and then began to subside again.[1] By 1960 it was also clear that the incidence among these survivors of cancers of the thyroid, breast, and numerous other organs, including the intestines, stomach, bone, lung, and ovaries, had risen to many times their levels in comparable, nonexposed populations. Over three-quarters of the women who were pregnant at the time of exposure to these bombs suffered some kind of reproductive or teratogenic effect: 25 percent gave birth to children with defects, especially abnormally small heads and mental retardation, 25 percent bore infants that did not survive their first year, and 28 percent miscarried.[2] There is no evidence to date of increased genetic disorders in the surviving descendants of this group of exposed women, but it is far too early to tell whether or not such effects will eventually appear.

The story of radiation research and its use in the early years of this century is strewn with endless examples of how a lack of appreciation for the hazards of chronic exposure to radiation could lead to tragedy. Within a decade of Roentgen's discovery of X rays in 1895 the first case of cancer caused by excessive X-ray exposure had been reported in medical journals. Marie Curie, the discoverer of radium, succumbed to leukemia, as did her daughter, undoubtedly because of the radiation exposures their work involved. Radiologists testing their equipment day after day by exposing their hands to it, dentists holding X-ray film in place in their patients' mouths with their fingers, industrial fluoroscopists checking for cracks and other flaws in metals with primitive, badly focused, and inadequately shielded X-ray equipment were among the early casualties of the widespread ignorance of radiation hazards.

In many cases the hands were the main target of these excessive exposures, and the effects on them ran the gamut from simple "cracking and roughening" of the skin to blistering, "ulceration and gangrene" causing "intense pain, scar formation, and deformity."[3] Certain types of exposures had a pronounced tendency to become malignant, first producing a skin burn that:

> . . . instead of disappearing in a few weeks, became an ulcer that penetrated to deeper layers of the skin and to the subcutaneous tissues. Here the process was exceedingly slow and obstinate and had an almost malignant tendency to progress and resist treatment.
>
> As time passed . . . it was learned that the skin lesions of the slowly penetrating variety might go on to loss of deeper tissues from ulceration, necessitating amputation of portions of the fingers and sometimes repeated amputations until only a stump of a hand or arm was left. It also became evident that X rays were capable of so altering the tissues that the ulcers underwent malignant transformation. Such X-ray cancers were found to be malignant, and if not promptly treated, they metastasized [spread to other parts of the body] early. A striking feature was the long latency between the original severe lesions and the development of the malignant transformation in the persistent ulcer.[4]

THE RADIUM DIAL PAINTERS' CASE

In the early 1920s dentists practicing in the area of Essex County, New Jersey, began noticing that many of their young female patients were developing strange complications following otherwise routine dental procedures. In one case, for example, a simple tooth extraction led to ulcers on the inside of the patient's cheek, anemia, and inflammation and infection of the bones of her jaw so severe they were eventually reduced to the consistency of jelly. Physicians called in to help treat these mysterious cases were at a loss to understand them, misdiagnosing them for some time as rheumatism, syphilis, or even an esoteric bone disorder called "phossy jaw," caused by exposure to phosphorus, that commonly afflicted workers in match factories. By 1924 approximately fifty cases, nine of them fatal, had been identified in this one small area, and public alarm was mounting.

Finally, in that same year an inspired bit of medical detective work by New York dentist Theodore Blum revealed that all of the women involved had at one time or another been employed in a local factory as painters of luminous dials on clocks and watches, using a paint made from water, zinc sulfide, manganese, copper, cadmium, acacia, and two radioactive substances, radium and mesothorium, both alpha-emitters. At this one factory and almost no others, dial painters had developed the habit of keeping a sharp point on their camels' hair brushes by drawing them through their lips. In doing so they inadvertently swallowed small amounts of the radium and mesothorium in the paint. During an average work week, investigators later estimated, the typical dial painter ingested anywhere from 15 to 215 micrograms of these two substances.[5] Symptoms appeared only in those employed for at least one year, and often lay dormant for several years after they had left the job, making the connection with their occupational exposures exceptionally elusive.

Though the practice of pointing the brushes by mouth was, of course, halted as soon as it was identified as the cause of these disorders, the casualty count continued to rise. By 1928 at least two victims had developed bone cancer, which in one case was the immediate cause of death. In others, crippling but nonmalignant bone damage developed in weight-bearing parts of the skeleton—bones of the feet, legs, arms, shoulders, and spine. The factory itself was found to be extensively contaminated with radioactive dusts and gases, and at least two employees who were not dial painters died of aplastic anemia caused by inhalation and possibly skin absorption of the radioactive substances at large in it.*

By 1929 the tragedy had claimed fifteen lives. In most cases the immediate cause of death was severe anemia, a loss of red blood cells that can lead to numerous kinds of dysfunctions in the body. Several autopsies confirmed the presence of radioactive residues, primarily mesothorium, in victims' skeletons, in amounts ranging from 50 to 150 micrograms.[6] Today, the consensus among radiation specialists

*As we mentioned in Chapter 4, aplastic anemia is a nonmalignant disease of the bone marrow that causes decreases in red and white blood cells and platelets.

is that as little as 1.2 to 2 micrograms of these substances deposited in the bones of the body may constitute a fatal dose.

Mesothorium, the main contaminant in these cases, has a physical half-life of just under seven years. Radium, however, though it never constituted more than 30 percent of the radioactive residues in the bodies of these women, has a half-life of some seventeen hundred years, and has continued to cause fatal bone cancers in some members of this unfortunate group as recently as the late 1960s.*

URANIUM MINING, LUNG CANCER, AND BIRTH DEFECTS

As the radium dial painters' tragedy demonstrates, ingestion of alpha particles can be quite hazardous. Inhalation of alpha-emitters is also harmful and is probably more common, since radioactive elements distributed widely in the earth's crust constantly emit radioactive gases as they decay. Among the more abundant radioactive elements in the soil, including carbon-14, potassium-40, and thorium-232, is uranium-238, the raw material for the fuel used in commercial nuclear reactors. Uranium ore is mined extensively in the West and Southwest, particularly in the Four Corners region where Arizona, New Mexico, Colorado, and Utah abut.

Among the many products of natural uranium decay are radium and radon, an alpha-emitting gas. Under most conditions radon, which has a half-life of about four days, is diluted and dispersed by the natural movement of air currents across the earth's surface. But in poorly ventilated mine-shafts—and in some very tightly insulated homes and other structures—it can accumulate to concentrations that may pose serious health risks.

By 1950, shortly after uranium mining had begun in earnest in the Southwest, radon was already strongly suspected as a possible cause of lung cancer among workers in these mines. But since the disease has a latency period of some fifteen years or more it was not until the mid-1960s that grim confirmation of this suspicion was finally forthcoming: In the past two decades the incidence of lung cancer among uranium miners has peaked at about six times the rate in similar nonexposed groups. As with asbestos workers, uranium

*Will your luminous watch dial give you a dangerous dose of radioactivity? No, not unless it's quite old, and even then the danger is minimal. New watches use tritium, not radium or mesothorium, and their small levels of beta emissions are almost completely blocked by the watch case. Even if they weren't, your dose from this source wouldn't even reach 1 mR per year.

miners who are also cigarette smokers are at even greater risk—ten times that of their nonsmoking coworkers. Fortunately, adequate protection from the dangers of radon gas inhalation in mines can be easily provided by standard ventilation equipment.

Another radioactive element, thorium-232, also decays into a potentially harmful radioactive gas, thoron. But since thorium is invariably strip-mined rather than dug out from deep underground shafts, there is normally plenty of natural dissipation of the gas by wind currents on the earth's surface and comparatively little hazard to thorium miners.

As we noted in Chapter 4, in highly insulated structures built over natural deposits of uranium or thorium, or constructed of concrete, bricks, granite, or other masonry material containing these elements, radon and its decay byproducts can accumulate to potentially hazardous levels. A series of studies conducted throughout the 1970s found that indoor radon levels in most houses ranged from a low of 0.05 picocuries* per liter of air (pCi/L) to a high of 18 pCi/L,[7] compared to an outdoor average of 0.05 pCi/L to 0.13 pCi/L. As with mining shafts, the key to preventing this problem is sufficient ventilation. Currently, air quality experts agree that any ventilation system providing a complete exchange of all the air inside a structure at least once every two hours offers adequate protection against all indoor pollutants, including radon. Although there is no formal regulation or standard governing indoor radon concentrations, most available evidence indicates that it should be kept below 3 pCi/L.[8]

Uranium mining has recently been linked with yet another type of health hazard, one that affects the public as much as the men who work in the mines and processing plants. Recent studies of birth records in the Four Corners area strongly indicate that ever since the uranium boom took off there a few decades ago, families in that region have been experiencing a disturbing rise in reproductive disorders, childhood cancer rates, and birth defects. In certain counties the incidence of some types of birth defects is three times as high as national levels.

Investigators believe that radiation may be to blame and that its major source in the region is the enormous mounds of waste tailings generated over the past thirty years by uranium mining and processing plants. Traditionally regarded as harmless rubble, tailing mounds were treated with lethal casualness: Children were allowed to play on them, and in at least one notorious case, not far from the

*A curie is approximately the amount of radioactivity emitted by one gram of radium. A picocurie (pCi) is one-trillionth of this amount.

Four Corners area, in the town of Grand Junction, Colorado, local contractors exploiting what they thought was a bonanza of cheap landfill and aggregate for making concrete used over 270,000 metric tons of tailings in making roadbeds and in the construction of thousands of buildings of all kinds, including homes, churches, hospitals, and schools.

Recently, careful measurements of radiation exposures caused by tailings in Grand Junction and near Salt Lake City, Utah, revealed that people living within a half mile downwind of the heaps were receiving exposures ranging up to 8,000 mR per year in the first location and nearly double that in the second. By comparison, the maximum exposure level for the general public allowed by various radiation regulatory agencies in this country is only 500 mR per year, exclusive of medical X rays or other health care sources of radiation. According to one recent report, one in every ten of the residents living in these buildings has been receiving an annual radiation dose equal to over five hundred chest X rays as a result.[9]

Since radon and thoron can also dissolve into groundwater, drinking contaminated water or boiling it may expose the stomach or the lungs to alpha-emitting radioactivity. Researchers speculate that in some locations of the country elevated levels of stomach and lung cancer may be linked to high concentrations of these gases in the water supply, and recommend stringent testing for their presence. As we noted in Chapter 5, since 1974 the Safe Drinking Water Act has limited the allowable level of alpha particle activity in water to 15 picocuries per liter.

MAJOR SOURCES OF RADIATION EXPOSURE

NATURAL

All of us are constantly exposed to low levels of "background" radiation from cosmic rays, radioactive compounds in the soil and the residues they emit into the air, water, and food crops, and from masonry building materials. Even the very cells of our bodies contain some natural radiation, in the form of potassium-40; most of us probably receive more internal radiation from this source than any other, under normal conditions.

Although levels of background radiation do exceed 1,000 mR per year in a few parts of the world, the average American is only exposed to about 100 to 125 mR/year at sea level. However, because the thinning atmosphere provides less and less protection against cosmic rays at higher altitudes, exposures increase with increasing elevation, tripling every three thousand meters or so. For this reason residents of states like Colorado, Utah, Wyoming, and New Mexico receive significantly more natural radiation from cosmic rays than residents of other states. The average Denver resident, for instance, has about twice the amount of cosmic radiation exposure each year that a person living in Honolulu does, simply because of the difference in elevation between the two cities. And the crews of jet aircraft receive even higher exposures, high enough to make it necessary for them to limit the amount of time they spend above thirty thousand feet each year.*

Background radiation probably does contribute in a small way to cancer, birth defects, and other health disorders in the general population, but the extent of this contribution is impossible to measure in any meaningful way because of the presence of so many other causes of disease in the environment, and because, compared to other sources of these kinds of disorders, the percentage caused by exposure to background radiation is quite small.

MAN-MADE

Health Care

Since the advent of X-ray technology the average American's level of radiation exposure has risen by 50 to 90 mR/year, or more than 50 percent of his or her average annual exposure, because of natural background levels. Over half of us are exposed each year to some type of medical or dental diagnostic X ray. It is extremely difficult to generalize about radiation doses caused by X rays, since different parts of the body can pick up different doses from the same X-ray exposure. However, we can say that a single chest X ray increases our gonadal exposure by some 5 to 8 millirads, a dental X ray by 9 millirads or so, while a full radiological examination of the GI tract may involve exposures as high as 1,700 millirads, and X rays of the lower back may run as high as 2,500 millirads. Most of these exams also expose other parts of the body, to varying degrees. The most hazardous X ray of all is of the abdomen of a pregnant woman, and is undertaken only in the most urgent circumstances.

There are other medical uses of radiation besides diagnostic X rays, many of which involve much higher exposures than the aver-

*The legal maximum exposure level for radiation workers—a classification that includes jet aircraft crews—in this country is 5,000 mR/year.

age X ray causes. For example, plutonium-powered pacemakers subject their users to annual exposures of about 5,000 millirads. Of course, in this case, as in every case where high radiation exposures are incurred in medical procedures, the benefits are judged to outweigh the risks. As we noted above, some types of cancer treatment involve exposures on the order of 4,000 to 5,000 *rads,* sufficient to cause burns and unpleasant side effects like nausea, but in many cases well worth these drawbacks.

Some years ago, before focusing and improved shielding were standard, health workers themselves were also routinely exposed to high, hazardous doses of radiation. As a number of studies of the causes of death among radiologists have noted, in the first decades (i.e., 1920 to 1949) of the profession radiologists suffered high rates of leukemia, certain types of cancer, and aplastic anemia, almost certainly as a result of chronic exposure to low doses of radiation in their work.[10] Today, the typical radiologist or radiological technician is probably exposed to no more than an extra 100 mR per year due to his or her occupation, and these excessive mortality rates have dropped to normal levels.

Medical and dental radiation constitutes by far the greatest source of man-made radiation exposure for most of us, a fact that has led most health care professionals to become increasingly cautious about ordering routine diagnostic X rays in recent years. Even so, there is little doubt that, as with natural radiation, we pay a small price for the benefits of medical radiation in terms of slight increases in the number of cases of cancer and birth defects that occur each year.

Recently, the Committee on the Biological Effects of Ionizing Radiation (known as the "BEIR" Committee) of the National Academy of Sciences-National Research Council attempted to estimate this price. For exposures of 1 rad per year the committee calculated that the normal lifetime cancer death rate in a representative American population of 1 million individuals (currently about 165,000 deaths) would be increased by 3 to 8 percent.[11] In other words, an annual radiation dose of about 1,000 mR above natural background levels will cause an additional five thousand to thirteen thousand cancer deaths per 1 million Americans. A controversy over this report has arisen because a minority faction of the BEIR Committee, including its chairman, believes that this estimate is too low, that it "underestimates the low-dose risk by a factor of 2 or more."[12] Remember, though, that most of us average only about 200 to 500 mR of radiation exposure per year, well below half of the 1 rad level used in the BEIR-III report, and much of that is not whole-body exposure.

Fallout

Since 1945 the desire of a growing body of nations to add nuclear weapons to their military arsenals has introduced a unique kind of radiation exposure to the category of man-made sources: fallout from the atmospheric detonation of weapons prototypes and components for testing purposes. Explosion of a nuclear device in the atmosphere releases an intense shower of many kinds of radiation, most of which fall back to earth in the immediate vicinity of the blast within hours. Exposure to the fallout in the area of a nuclear blast can cause a variety of health effects, including ARS and skin burns ranging from superficial to severe. Within 80 to 160 kilometers of the blast, alpha- and beta-emitters can contaminate air, water, and food supplies. Some fallout is also pulled aloft into the upper atmosphere by wind currents and carried around the earth for years, even decades, constantly subjecting populations in its path to a slow, steady drizzle of radioactivity.

Among the more hazardous substances released in these events are iodine-131, cesium-137, strontium-89 and 90, and plutonium-239. The shortest lived of these is iodine-131, with a half-life of only a bit more than 8 days. Iodine of all kinds has a strong affinity for the thyroid gland. As well as being a potential cause of thyroid cancer, iodine-131 irradiation of the thyroid gland of the human fetus is capable of causing stunted growth, mental retardation, and other birth defects. (Radioactive iodine is also used in medicine to treat thyroid cancer and hyperthyroidism, a use that carries with it some risk of causing leukemia since it can involve exposure of the bone marrow.)

Strontium-89, also relatively short lived, has a half-life of only some 50 days. Its close cousin strontium-90, with an effective half-life of some twenty-eight years in the human body, has a strong tendency to collect in the bone marrow, and is a potential cause of leukemia and bone cancer. Cesium-137, a long-lived alpha-emitter, has a half-life of some 30 years, but tends to gravitate to the less radiosensitive muscles. Plutonium-239, which we discuss in more detail below, has the longest half-life of any of these fallout byproducts, some 24,400 years, and is easily the most toxic radioactive compound known to man, one that is especially hazardous if inhaled. Plutonium tends to become deposited in and near the reproductive organs in the human body; most of us now carry a minuscule but detectable amount of plutonium in our bodies as a result of exposure to fallout.

In the late 1950s concern about these fallout byproducts and the fact that some were being routinely found in the food supply, especially in milk and vegetables, led to the signing of an international treaty banning atmospheric tests of nuclear devices by Great Britain, the United States, and the USSR in 1962. At that time the average annual radiation dose from fallout in the United States was approaching 12 mR. By the end of the decade it had dropped to one-third that level, mainly due to the ban on atmospheric tests, which in this country were replaced by underground detonations. However, almost as soon as the ink on the test ban treaty was dry, other nations began to qualify for admission to the nuclear club, beginning with France in the mid-1960s. Since then there has been a trickle of new members, the most recent one being China. Since many of these countries have seen no reason not to go ahead with atmospheric test programs of their own, the radiation exposure level in the world due to fallout is *slowly* beginning to climb again, heading toward a predicted level of about 5 mR per year per person by the year 2000. As a trend this is clearly undesirable, but the exposure levels involved are not especially high: 5 mR per year is approximately the same radiation dose the average American now receives from electronic consumer products (e.g., color TVs and smoke detectors) that emit some kind of radiation.

A few years ago we discovered that we had not escaped unscathed from the years of atmospheric testing that took place before the 1962 ban was signed. In 1979, for example, researchers reported that the leukemia rate among children of a small town in Utah named St. George had been abnormally high—some 2.5 times the national average—for a number of years following the ban. The town's cancer rates were also suspiciously elevated. The cause, as usual, was not entirely clear, but the fact that St. George is only a few miles downwind of a nuclear test site at Yucca Flat, Nevada, where we conducted much of our preban atmospheric testing, seems to be more than just coincidence. There is even a strong supposition that an unusually high incidence of fatal cancers among the members of a movie cast—including John Wayne, Susan Hayward, and Dick Powell—who filmed parts of a picture called *The Conqueror* in 1954 in a nearby canyon may have been due to excessive fallout exposure.

Nuclear Power

The first commercial nuclear reactor built for the purpose of generating electrical power began operating in this country in 1958. By 1981 seventy-three commercial reactors all across the country were generating slightly more than 10 percent of our electricity. Another eighty-four reactors were already sited and ready to be built, and

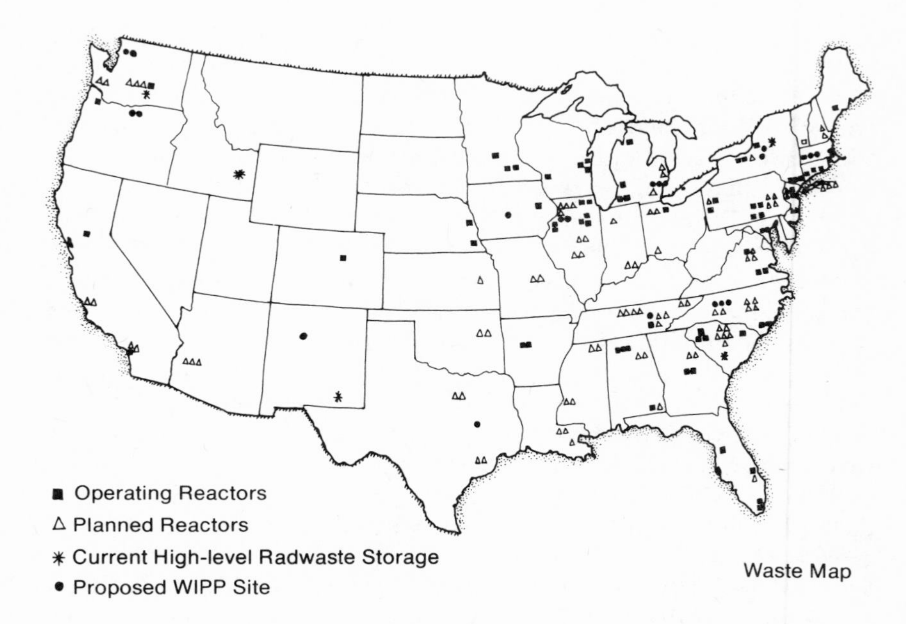

Figure 23 A

thirty more were in various stages of the planning process. As of this writing the Reagan administration had made it clear that it planned to do all it could to promote nuclear power.

Older reactors were designed to produce slightly less than 1,000 megawatts each—an average of about 750 megawatts per reactor.* The next generation of reactors will be bigger, in terms of power output, generating an average of just over 1,000 megawatts each. The typical 1,000-megawatt reactor has an operating life of some twenty to thirty years, steadily becoming more and more radioactive with each passing year until it finally becomes too hot to handle safely and must be permanently shut down and either sealed off or disassembled and disposed of. Certain components of a mothballed or disassembled reactor's core remain highly radioactive and extremely hazardous for hundreds of thousands of years.

The nuclear fuel cycle begins with the mining, extraction, purification, and other processing steps of uranium ore, the health hazards of which we touched on above. At this stage processed uranium is predominantly (over 99 percent) a variety known as U-238. It also contains a small amount, less than 1 percent, of relatively scarce U-

*A megawatt, 1 million watts, is equal to 1,000 kilowatts.

235. For efficient fission the percentage of U-235 must be brought up to 3 to 4 percent, a step known as "enrichment." (For some newer types of reactors and for some military applications the amount of U-235 must be brought up to 90 percent or more.)

Increasing the proportion of U-235 in a material can be accomplished in a variety of ways. One method, based on the fact that U-235 has a mass different from that of other substances, uses a magnetic field to pull U-235 particles into a distinct orbit, from which they can be collected. Other methods depend on the fact that heat will cause U-235 particles to travel at a different velocity from other particles, thus making possible their separation from complex mixtures and reconcentration at any desired proportion. Uranium is initially extracted from its surrounding ore in a standard metallurgical fashion, by dissolving the ore in an acid bath, which renders the uranium metal soluble, transforming it into a gas, and finally reconverting it into a purified solid.

Enriched uranium is then fashioned into 2.5-x-1.2-cm cylindrical pellets, which are inserted into long (3.5 to 4 meters), thin fuel rods made of a special alloy of zirconium. In a typical reactor some thirty thousand to forty thousand of these rods are then inserted into a steel vessel, the heart of the reactor core, which during operation is always kept filled with water or another coolant.

Power is generated in the core because the unstable atoms of the enriched uranium fuel have a natural tendency to disintegrate. As they do, they emit intense radioactivity of all kinds, including neutrons. When neutrons strike other uranium atoms in the core, they cause new disintegrations, the emission of still more neutrons, and so on. When this "chain reaction" becomes self-sustaining, the reactor is said to have "gone critical," and is ready to generate power.

As well as releasing hundreds of highly radioactive byproducts, fission also produces intense heat, and this is what is used to create electricity. A gas or fluid, usually water, is circulated around the reactor to collect some of this heat, which is then carried to equipment that uses it to boil water into steam. The steam is then employed to drive the turbines that actually produce the electrical current. It takes 30 trillion fissioned uranium atoms to generate 1 kilowatt of electrical power.

The fission process is controlled by another set of rods, called control rods, which are coated with a substance like cadmium or boron that absorbs neutrons. When the control rods are extended into the core, they capture so many neutrons that a chain reaction becomes impossible to maintain. If they are slowly withdrawn, more and more neutrons strike home and fission occurs at ever more intense levels; this is in fact how fission is initiated when a reactor is ready to be started up.

The main danger posed by commercial reactors is release of the intense radioactivity generated by fission into the ambient environment in the form of gases, contaminated steam, or radioactive water. Commercial power reactors cannot explode like an atomic bomb. The worst thing they can do, if their cooling systems malfunction, is become so hot that the highly radioactive core could theoretically melt through the floor of the building it is housed in.

To minimize the risk of radioactivity escaping from power reactors, their fuel and control rods are surrounded by a steel reactor vessel with walls several inches thick. Around the vessel are many feet of concrete shielding. The entire reactor core is housed in a "containment" building, which is made of concrete walls that are many feet thick. The containment building is also designed so that it can be sealed off from the outside environment in an emergency.

ROUTINE EMISSIONS. Nuclear reactors routinely release small amounts of radioactive gases, primarily xenon, krypton-85, and radioactive iodine into the air, and flush small amounts of liquids contaminated with radioactive tritium and cesium into local waterways. Under normal operating conditions NRC regulations prohibit any facility in the nuclear fuel cycle, including reactors, from emitting radioactivity that will expose individuals living closest to the plant to more than 25 mR per year, far less than the amount each one of us receives annually from natural background sources of radiation. Based on careful studies of the environment around routinely operating reactors, experts feel that the actual amount of additional radiation exposure received by those living closest to them is much less than this allowable level, probably well below 5 mR per year.

However, the more important issue is that, like any other human construction, nuclear power plants do not always run routinely. They break down, catch fire, and fail almost daily, in countless trivial and some not-so-trivial ways. Most of the mishaps and human blunders that take place in reactors pose no more serious a threat to human health than the average accident in any other industry, and most are quickly corrected. But every once in a while, more often than we might like, they pose the extremely grave possibility of one of the worst kinds of environmental disasters conceivable—a meltdown of the highly radioactive reactor core—and this is where the hazards of nuclear power depart in a radical way from those of every other industry.

MELTDOWN. The heat generated by fission is so intense that, if not carefully controlled, it would soon melt the contents of the core and send tons of molten slag through the floor of the containment building. The fanciful possibility that this extremely radioactive lump of hot (over 2,875°C) metal would continue burning a hole straight through the Earth has given rise to the name "China syndrome."

A full meltdown would be an unparalleled environmental catastrophe, releasing high levels of radioactivity into the air, soil, and both surface and groundwater supplies. A plume of radioactive gases and steam would rise into the sky, to be carried wherever the wind blew. Whatever was touched—buildings, people, plants, animals—would become contaminated. Most of the worst damage would occur within a 15-kilometer radius of the stricken reactor, but for as much as 150 kilometers or more downwind anyone who inhaled contaminated air, ate contaminated food, or even absorbed fallout through the skin would run the risk of developing radiation sickness, depending on the dose of the exposure. A large area, perhaps hundreds of square kilometers, around the plant itself would be contaminated for years, decades, or even centuries to come.

Even the extremely remote possibility of such a disaster has compelled nuclear reactor designers to provide numerous safety systems in commercial power reactors. First of all, they are not sited in heavily populated areas, and they are heavily shielded with concrete and steel. Every reactor is equipped with not only a primary cooling system, usually containing water, but also backup circuitry and redundant pumps, as well as an Emergency Core Cooling System (ECCS), capable of pumping thousands of liters of water a minute into the core should the primary and backup systems fail. There is also a "scram" system, an automatic mechanism that causes the control rods to drop, fully extended, into the core, shutting down all fission. The containment building that houses the reactor is kept under constant negative pressure so that if leaks develop, the outside atmosphere will tend to leak in rather than vice versa.

But all of these systems are dependent on mechanical devices, electrical circuits, effective monitoring and warning systems, and human beings, none of which are notorious for operating perfectly. Until the Three-Mile Island accident some of these devices had never been fully tested under actual emergency conditions, and as we shall

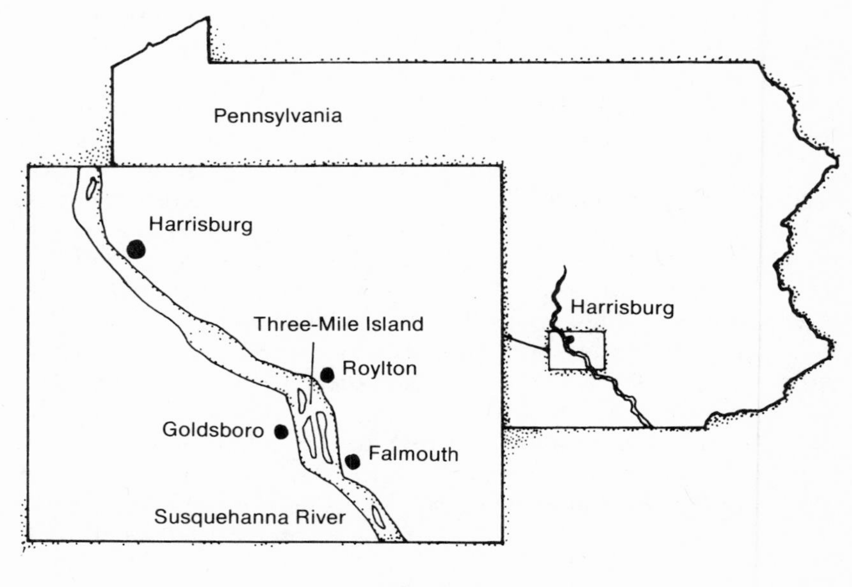

Figure 24

see below, when put to the test some proved to be less than fool-proof.

THREE-MILE ISLAND. A few miles southeast of Harrisburg, Pennsylvania, on an island in the Susquehanna River, sit two nuclear reactors owned by a consortium of local utility companies and operated by the Metropolitan Edison Company. Together, the two reactors (usually designated TMI-1 and TMI-2) were until March 1979 capable of generating about 1,700 megawatts of electricity.

In the early morning hours of Wednesday, March 28, 1979, TMI-2 was running at nearly 100 percent of its peak generating capacity when, just a few seconds after four A.M., the main pumps that supply water to the steam generation system stopped working. Automatically alerted to the imminent loss of steam this would cause, the turbines that generated TMI-2's electricity halted a few seconds later.

In a reactor like TMI-2, water that feeds into the steam generators does not mix directly with the water used to cool the reactor, which circulates in a completely separate piping system. Since water kept under pressure boils at a higher temperature than unpressurized water, coolant water in many reactors, including TMI-2, is pressurized so that it won't instantly vaporize as it circulates around the hot core.

To shed some of the heat it picks up from the core, coolant water is routed close to steam-generating water (known as "feedwater"), which then absorbs some of the heat from the coolant water and supplies it to the steam generators. If the feedwater stops circulating, the coolant water has no place to dump its steadily accumulating heat load and its temperature quickly begins to rise. As its temperature rises, it expands, pressure inside the pipes that carry it begins to build up, and eventually relief valves must open to bleed off some of this pressure or seals and pipes will begin to rupture. If for some reason the relief valves don't close again, enough pressure can be lost in the system so that the coolant water does begin to flash into steam. Since steam cannot cool as well as water can, the reactor itself will begin to heat up. This is essentially what happened at TMI-2.

As the main feedwater pumps stopped operating, TMI-2's coolant water temperature and pressure both began to rise. The pressure quickly built up to about 2,250 pounds per square inch, 100 psi above normal. At this point a relief valve in the coolant line popped open to relieve the pressure buildup. It was supposed to close again automatically when the pressure dropped below a certain point, but it didn't. It may have cycled open and shut a number of times, but it finally remained stuck in the open position. At the same time, however, indicators in the central control room panel appeared to show that it had closed normally. For the next two hours and twenty minutes it stayed open, undetected, allowing precious coolant water to spill out onto the floor of the containment building as the pressure dropped and the water level around the core sank lower and lower. If the operators of TMI-2 hadn't finally figured out what was wrong, and brought water in from emergency sources, the core might have melted and a large section of southern Pennsylvania would be radioactive wasteland today.

Within seconds after the main feedwater pumps first stopped at four o'clock, the reactor at TMI-2 automatically scrammed. Fission came to an abrupt halt, but there was still enough residual heat in the core to cause problems. Then, because its relief valve was stuck open, the pressure level in the coolant system started to drop.

For the next few hours, until operators finally realized that this valve was stuck open, they were misled by a monitor in the coolant system's pressurizer tank that seemed to show that water levels in the system were high—normally a reliable sign that there is sufficient water in the system to maintain the right pressure. Operators weren't sure what to believe: Indicators showing that coolant system pressure was dropping or monitors showing that there was plenty of water in the coolant system. For nearly two and one-half hours, faced with this dilemma, they made the wrong choice.

Ordinarily in a situation like this one at a nuclear power plant it wouldn't matter so much what the operators thought because the ECCS pumps are designed to come on automatically when the pressure in the cooling system drops below critical levels. And, just as they were meant to, the ECCS pumps at TMI-2 came on within minutes after the main feedwater pumps quit working just after four A.M. But, in their second serious misjudgment of that fateful morning, TMI-2 operators, still thinking that there was plenty of water left in the coolant system, *turned one of the two ECCS pumps off and reduced the flow from the other one to a mere trickle,* thus crippling the very backup system intended to prevent the catastrophe that almost occurred.

Within six minutes after the accident first began, pressure in the coolant system dropped so low that water around the core began flashing into steam, pushing water in the system back and causing the water level in the pressurizer tanks to give false readings, thus compounding the operators' mistaken assumption that there was sufficient water left in the coolant system as a whole. At this point TMI-2 operators made a third critical error in judgment, deciding to *remove* even more water from what they assumed was a dangerously full—rather than half-empty—coolant system. To do this they began transferring it from the containment building to an adjacent "auxiliary" building. Since the piping system between the two buildings was not airtight, it released small amounts of radioactive gases that had dissolved in the coolant water out into the ambient environment. To make matters even more confusing, within ten minutes after the main feedwater pumps had first quit working, water spilling from the stuck relief valve began flooding onto the containment building floor. It, too, was piped into the auxiliary building, which became increasingly radioactive.

At this point, about an hour and a half after the accident first began, the coolant system pumps, never built to circulate steam instead of water, began to vibrate heavily, so the operators shut them down, compounding all their previous mistakes. Now there was not only a rapidly diminishing amount of water in the coolant system, but the circulation of what there was could not be readily controlled, adding to the likelihood that water around the core would continue to vaporize.

As the temperature in the core began to rise, its fuel rods began to expand and rupture, releasing radioactive gases into the containment building. By six o'clock that morning, two hours after the accident began, radiation monitors in the containment building were sending constant alarms to the TMI-1 control room, where operators had set up a command post and were working frantically to bring TMI-2 under control again. At the same time, though no one in the control room had yet realized it, the zirconium alloy used to make the fuel rods was reacting with steam to produce hydrogen gas, which is explosive. Furthermore, the bubble of hydrogen gas that began to form at the top of the reactor vessel also helped block what water there still was in the coolant system from circulating around the core. By six fifteen the top of the core was exposed. Half an hour later over half the core was uncovered and its temperature was approaching 2,200°C, only some 650°C short of the level at which a meltdown would have occurred. At seven o'clock TMI's managers declared a site emergency and alerted local public safety officials and federal regulatory agencies to the problems they were having bringing the reactor under control. Twenty minutes later a monitor at the top of the containment building was recording radiation levels as high as 800 rems per hour. Along the edge of the island, however, radiation readings were still low, less than 1 mR per hour.

Finally, at seven twenty, operators turned the ECCS pumps back on briefly, the single move that averted a major catastrophe. After a great deal of confusion, the fact that gauges continued to show large amounts of water were still escaping into the containment building led operators to conclude that the valve was stuck open, a deduction that was later confirmed by the discovery of a warning light that had been obscured under a maintenance tag. Shortly before seven thirty, still unsure of their control over the reactor, TMI officials upgraded their site emergency to a general emergency, one involving a potential threat to public health.

A local radio station first broadcast the story of the TMI accident at about eight thirty Wednesday morning. Within an hour the White House had learned of the situation. By late that afternoon the whole country was watching TMI with bated breath. Shortly before noon that day, operators in the TMI-1 control room heard a small thud in the containment building—an explosion of accumulated hydrogen gas—but didn't understand what it was or the danger it implied.

Efforts to bring the pressure in the coolant system under control continued without success. As operators worked to add or remove water from the cooling system, pipes running between the containment building and the auxiliary building continued to leak small amounts of radioactive gases into the outside air. By midafternoon, Thursday, March 29, radiation readings in the nearest towns were still at a low 1 to 2 mR/hour level. But helicopters monitoring emissions just above the containment building were beginning to pick up brief readings of high radiation levels—as high as 3,000 mR/hour.

Early Friday morning an intentional transfer of radioactive gases from one tank to another caused a release that a hovering helicopter read as 1,200 rems/hour (1.2 million mR/hour) in the air above the stricken reactor, a level that caused deep concern and alarm among local health and public safety officials. At twelve thirty P.M. that day Pennsylvania Governor Richard Thornburgh issued a public advisory recommending that pregnant women and preschool children evacuate the area within a five-mile radius of TMI and that all schools in that same area be closed until further notice.

By Friday afternoon, though the threat of a meltdown seemed to be past, it had been replaced with the danger of what TMI operators had finally recognized was the large bubble of hydrogen gas that had formed inside the containment building. Feverish efforts to devise ways to shrink or dissipate it continued for the next two days, until instrument readings on Sunday afternoon indicated that it had begun to disperse spontaneously. Gravely concerned about the TMI situation, President Carter visited the site that same afternoon. Over that weekend a massive HEW effort brought in some 237,000 bottles of potassium iodide, a drug that prevents radioactive iodine from reaching the thyroid gland, for possible distribution to local residents and TMI workers. Never used, the drug was later shipped to a federal storage depot in Arkansas.

Finally, by Monday morning, five days after it began, it looked as though the worst commercial nuclear accident in American history was over and TMI-2 was under control again. It would be some time before a "cold shutdown"—when the coolant temperature is so low that steam cannot form—was reached, but the crisis was over. On Wednesday, April 4, one week after the accident began, Governor Thornburgh rescinded his evacuation advisory.

Had the gases that regularly leaked from TMI-2 during the worst hours of the accident harmed anyone? Some six months after the accident the presidential commission formed to study it said no. Headed by John G. Kemeny, President of Dartmouth College, the Commission concluded that:

> The maximum estimated radiation dose received by any individual in the off-site general population (excluding the plant workers) during the accident was 70 mR. On the basis of present scientific knowledge, the radiation doses received by the general population as a result of exposure to the radioactivity released during the accident were so small that there will be no detectable additional cases of cancer, developmental abnormalities, or genetic ill-health as a consequence of the accident at TMI.[13]

Not surprisingly this finding was almost immediately challenged. Ernest J. Sternglass, professor of radiological physics at the University of Pittsburgh School of Medicine, has asserted[14] that there was indeed a clear and strong correlation between the path of windborne iodine-131 from TMI and increased infant mortality rates for the months of April to July 1979, in parts of the four states where the concentrations of windborne TMI gases had been at their highest. Sternglass makes a vigorous case for his views, but to date they have not been supported by data on infant mortality rates in the immediate vicinity of the plant. After reviewing the data and determining that the infant mortality within a ten-mile radius of TMI was 19.3/1,000 for the entire period of January to June 1979—meaning that it was precisely the same immediately after the accident as before—Dr. George Tokuhata, Director of the Bureau of Epidemiological Research for the Pennsylvania Department of Health concluded that "there is no evidence to date that radiation from the nuclear power plant influenced the rise or fall of [death] statistics."[15] However, it is worth noting that neither the Kemeny Commission nor Dr. Tokuhata was prepared to rule out the possibility that TMI's radioactive emissions may have caused subtle, undetectable damage to infants, such as mental retardation or birth defects.

The danger of a meltdown at TMI-2 was over in a matter of days, but the plant is still hazardous as of this writing and will be for some time to come. There is no danger of the core heating up again, but the accident left the containment and auxiliary buildings filled with some 3 to 4 million liters of highly radioactive water, which will take years to decontaminate using special resin filters. The longer the contaminated water sits inside the reactor building, which is not watertight, the greater grows the risk of leakage into the ambient local soil and water supplies. Some 1.9 million liters have already been successfully treated in this way, but concern is now growing that the filters themselves, which are stored inside steel containers placed inside concrete vaults when they become too radioactive to be used anymore, may have become so radioactive that they are corroding the walls of these containers.

Once treated, though it will contain only a small degree of residual radioactivity, water from these buildings cannot be returned to natural waterways but must be disposed of by other means. Suggestions have included mixing it with concrete to form massive blocks that could be disposed of in conventional ways, such as by burial in landfill projects or by transferring it into large, open-air lagoons built on Three-Mile Island itself and allowing it to gradually evaporate into the atmosphere. The Metropolitan Edison Company had considered dumping it into the Susquehanna, but congressional action in mid-1981 forestalled this option.

The accident also left a sizable amount of krypton-85 gas trapped in the containment building, which was vented through special filters into the outside air from June 28 to July 11, 1980, without apparent harm. Air and groundwater sampling has been conducted more or less continuously since the accident by the EPA, and effluent monitors at the site have been set to sound an alarm any time radioactivity exceeds certain levels.

Getting rid of the contaminated water may be the largest problem TMI cleanup crews face, but it is not the most hazardous. The 36,816 partly damaged, highly radioactive fuel rods, weighing a total of some ninety metric tons, must also be removed and disposed of in one of the three sites available for high-level "radwaste," in Washington, New Mexico, or Idaho. The reactor building itself must either be sealed off and "mothballed" or meticulously disassembled, piece by piece, by heavily shielded workers, in what will be the largest "decommissioning" of a nuclear reactor ever attempted.

Estimates of the length of time it will take to clean up TMI-2 range from six to nine years, placing the final completion date somewhere around 1988. As of this writing, some $300 million has been spent on the cleanup process, and the total cost will run well over $1 billion, perhaps as much as $3 billion. According to a plan developed by Governor Thornburgh and endorsed by President Reagan in late 1981, the remaining expenses of the cleanup will probably be shared by TMI's insurers, other nuclear utilities, Pennsylvania and New Jersey, the utility company that owns TMI, and the federal government, which has committed itself to bearing more than $100 million of these costs.

RADWASTES

Every time a hospital technician snaps another X ray, or a university researcher spills a radioactive solution on a lab bench and then wipes it up, or a technician at a commercial reactor discards a pair of contaminated gloves, radioactive wastes—radwastes, for short—are produced. Aside from mill tailings, the greatest volume of radwaste generated in this country is what has in the past been referred to as "low-level" waste—a category that primarily includes things like old paper, tools, clothing, animal remains, contaminated soil, radiopharmaceuticals, and so on. Low-level wastes are usually bulky, have short half-lives (less than thirty years), and contain relatively little uranium or plutonium. They are most commonly disposed of by burial at special sites, but some may also be quite legally incinerated, flushed into municipal sewer systems, or simply tossed into the nearest garbage can.

Though low-level wastes generally pose a lesser hazard than other categories that we discuss shortly, regulations concerning their disposal have, on the whole, grown more stringent over the last few decades. For example, until Congress halted the practice in 1972, the Atomic Energy Commission (now superseded by the Nuclear Regulatory Commission and the Department of Energy) as well as the military services routinely dumped barrels packed with low-level wastes into both the Atlantic and the Pacific. Some thirty-three thousand barrels were dumped off Maryland and Delaware from 1946 to 1970, and forty-five thousand drums crammed with these same kinds of wastes were deep-sixed into a one thousand-fathom site near the Farallon Islands, only fifty kilometers or so off the coast of San Francisco, during this same period. No attempt was made to use containers sturdy enough to withstand saltwater corrosion or the water pressures at these depths, and recent surveys of the Pacific site have found that as many as one-third of these barrels are now leaking. Samples of sponges, sediments from the ocean floor, and edible fish taken from the area show radiation levels in excess of those expectable owing to natural or such man-made sources as fallout.

In distinct contrast to the real but fairly manageable problems posed by low-level radwastes are the disposal issues raised by "high-level" radwastes, primarily spent fuel rods contaminated with plutonium-239, 240, and other long-lived, extremely radioactive fission byproducts. According to some authorities a typical 1,000-megawatt reactor produces some 100 to 225 kilograms of plutonium a year, more than enough to give every human being in the world a fatal dose of lung cancer if evenly distributed and inhaled—a practical impossibility. (In fact, less than a quarter of a kilogram of plutonium is sufficient to wipe out the entire population of the United States, given these same *theoretical* conditions of uniform exposure.)

For many years we had an efficient way of getting rid of the plutonium that built up in old fuel rods. Used fuel rods were bathed in an acid solution that broke them down into their component elements, most of the reusable uranium and plutonium was separated out, and the remaining liquid wastes were then stored in large underground steel and concrete tanks constructed in the states of Washington and South Carolina for just this purpose. Although plutonium can be used to fuel a nuclear reactor, most recovered plutonium went into making weapons. (In fact, the bomb dropped over Nagasaki contained about four to five kilograms of plutonium.) But fears that a terrorist group might steal some of this reprocessed plutonium and use it to construct an outlaw bomb led the Carter administration to order a halt to civilian plutonium reprocessing in 1977. Only the Department of Energy was allowed to continue to reprocess plutonium, for our military needs, in a few well-guarded installations. President Reagan rescinded Carter's order in 1981.

The search for a solution to the problem of disposing of high-level radwastes has been going on for at least a decade, and has generated thousands of proposals ranging from the prosaic (simply collecting them and storing them in well-guarded bunkers until we can figure out how to destroy them for good) to the cosmic (former Energy Secretary James Schlesinger suggested loading them into rockets and firing them into the sun—a not entirely unreasonable idea except that our rocketry skills are not yet perfect enough to guarantee that they won't fail to escape the planet's gravitational pull and fall back, uncontrollable, to earth). At this writing the best-favored proposal seems to be burial deep in natural "salt domes" located in many of the Gulf Coast states.

Salt domes are huge underground deposits of salt left behind by the evaporation of prehistoric seas. Centuries of mining their salt has left vast underground caverns in them, extending thousands of meters into the earth. Disposal plans envision lowering canisters or "bricks" of radwastes hundreds of meters down to a cavern floor, where they would then be buried at carefully spaced intervals. The canisters or bricks would be composed of a tough, durable material, possibly a type of ceramic. Once the vault is full, it would be topped off with salt, sealed up, watched carefully for a few years, then left to its destiny.

Salt domes seem to have many natural properties that recommend them for use as radwaste dumps, or "National Geologic Repositories," as the federal government calls them. For one thing, the simple fact that they have been there for thousands of years argues that they are so geologically stable they will probably last a few more millennia. For another, they are relatively dry—a critical concern since water mixing with natural minerals in the earth can corrode the toughest man-made substances within centuries, and these wastes will still be toxic hundreds of thousands of years from now. Finally, salt deposits are probably impervious to the intense heat that will be generated for many decades to come by these wastes. Scientists doubt that they will burn, bulge, or explode, and if they develop cracks they will probably seal themselves up again naturally.

But all these points are so far just predictions, not proven facts. Opponents of the salt dome scheme aren't so sure that we can safely consign such poisons to even the most stable geological formation for a period that will last at least seven or eight times as long as recorded history without asking for trouble. To try to resolve some of these issues with more empirical data, the first actual test of this and other disposal methods was being initiated as this book was being written. At a site a few miles southeast of Carlsbad, New Mexico, the first federal Waste Isolation Pilot Plant (WIPP) had been authorized to begin testing permanent disposal methods for both low- and high-level radwastes, including a limited number of containers of spent fuel material. The NRC, which is in charge of the WIPP facility, has stated that its basic criteria for a permanent disposal method for high-level radwastes include containment for at least one thousand years without leakage. It will be at least another decade before results of WIPP tests indicate how feasible salt dome burial truly is, and it will be far, far in the future before our distant descendants learn the true consequences of this experiment.

REGULATING RADIOACTIVITY

There are at least four major federal radiation regulators, and almost as many advisory bodies with no formal regulatory powers but a great deal of influence over the policies and rules promulgated by official entities. The most important regulator of civilian nuclear power is the Nuclear Regulatory Commission (NRC), which took over much of the Atomic Energy Commission's functions in 1974. The NRC's basic mandate is to protect the health and safety of the public, to insure the integrity of the environment, and to safeguard nuclear installations from sabotage and theft. Five NRC commissioners and a staff of thirty-four hundred, working with a $500 million budget, are charged with licensing and inspecting all nonmilitary nuclear installations, including power-generating reactors and the fuel-processing facilities associated with them. Virtually all of the activities, emissions, and wastes associated with these facilities are under the NRC's supervision. (The Department of Energy oversees all military-related nuclear development, testing, and storage, including temporary waste disposal.) The NRC is also charged with developing and overseeing systems for the permanent disposal of the nation's high-level radwastes, from any source, civilian or military.

Critics have for years charged that the NRC is dominated by the nuclear power industry and serves as little more than a rubber stamp for its interests. In the wake of TMI, more serious charges have surfaced that if the NRC had paid more attention to its own accident-reporting system, it would have detected hazardous patterns of equipment failure, operator negligence and lack of training, and faulty maintenance practices—all foreshadowing the problems at TMI. Perhaps most questionable, according to these critics, is the NRC policy of leaving to nuclear utilities themselves the responsibility of conducting safety evaluations, with no more than an occasional spot check by NRC field inspectors to make sure that its guidelines are being followed.

Under the 1968 Radiation Control for Health and Safety Act, responsibility for the regulation of electronic equipment and consumer products—color TVs, video display terminals, lasers, smoke detectors, and so on—that can emit both ionizing and other kinds of radiation lies with the Bureau of Radiologic Health (BRH), a branch of the Food and Drug Administration. The BRH also oversees the use of radioactivity in health care fields, and conducts research and educational programs designed to reduce the hazards of X rays and other medical radiation exposures to both practitioners and patients alike.

Finally, the EPA's broad mandate to protect the nation's air, water, and soil includes the power to set limits for the amount of radioactive emission that can be allowed to enter the ambient biosphere from any nuclear facility. Under a variety of clean water, air.

waste disposal, and other recent laws, the EPA has acquired the authority to set general standards for radiation emission into the ambient environment. Other agencies, usually the NRC or DOE, are typically charged with implementing and enforcing these standards.

Under a program known as ERAMS (Environmental Radiation Ambient Monitoring System), the EPA is also responsible for monitoring fallout. ERAMS maintains at least one air-sampling station in nearly every state, and periodically collects and tests samples of water and, in conjunction with the FDA, pasteurized milk for analysis of their radioactivity content.

In September 1981 the Reagan administration made it clear that it intends actively to promote nuclear energy. Perhaps the risks involved in nuclear power and the myriad other peaceful applications of radioactivity that we have developed in this century are worth taking. Perhaps we will be able to manage them successfully, without having to learn the hard way about its destructive potential—waiting until we suffer a dramatic, massive public health catastrophe that kills or injures thousands or even millions of us.

However, the history of environmental health hazards does not seem to warrant this optimism. In their dealings with fossil fuels, pesticides, chemical wastes, and numerous other environmental health threats, industrial cultures have repeatedly demonstrated that the only thing that really provokes an effective response to large-scale health risks is the actual occurrence of all-too-predictable disasters, disasters that claim such a terrible environmental and human toll that they can no longer be ignored. Has Three-Mile Island served this purpose for us? Mitchell Rogovin, a lawyer who headed the NRC's own investigation into what happened at TMI-2, doesn't think so. Some two years after the accident, in response to a question about whether or not he thought the NRC had instituted better safety practices at commercial reactors to prevent future TMIs, Rogovin asserted that the NRC had not improved nuclear regulation at all. "No one would question whether there will be another accident," he is quoted as saying. "It's merely a matter of when."[16]

POSTSCRIPT: LOOKING AHEAD

Underlying all of the issues we've raised looms the same haunting question that we touched on in the first pages of this book: What will the future bring? What will our drive for progress cost us in terms of our environmental well-being? As new technologies introduce new hazards, will we be able to manage them? Or are we, in the final analysis, in danger of altering our environment in ways that are not only uncontrollable but irreversible?

At present these questions have different answers in different parts of the world. In the highly industrialized nations of the Northern Hemisphere the picture of the future, though sobering, is far from bleak. In these countries, as we have tried to point out, there are ample indications that the technical ingenuity needed to find solutions to problems of air, water, and food pollution, waste disposal, and occupational hazards is available. But it is also clear that the process of solving these problems is all too often a reactive one. Much too frequently it is not until human welfare is seriously threatened that a concerted effort is made to respond to pressing environmental health issues. In short, in the materially fortunate parts of the globe, the obstacles to sound environmental health policies tend to be matters of politics and economics rather than technology.

The one area where the developed nations of the world appear to face a truly technological dilemma is in our attempt to harness the atom to the generation of electrical power. *If* the problems of plant safety and radwaste disposal can be solved—and this is still a very big question—then we may ultimately find that nuclear power is our technological benefactor. If not, we will have to come to terms with the fact that this episode in our development is a dead end, and we will have to begin actively to turn our attention, time, and money to the cultivation of technologies—such as the science of transmuting sunlight directly into electricity via semiconductors—that hold the promise of making up for what we will lose by turning off the atom.

It seems clear to us that over the next two decades the single most critical factor influencing environmental health issues in the developed nations will continue to be energy costs and use patterns. For example, as we have seen over the last few years, increases in oil prices have had the power to force a massive consumer shift to smaller, usually imported cars. To remain competitive, our devastated domestic automobile industry is pursuing a shortsighted strategy, one that sacrifices environmental health—in the form of stringent emission standards—to economic advantage. This strikes us as a desperately myopic response for two reasons: First, it overlooks the fact that the public made it clear in a May 1981 Harris survey that, by a seven to one margin of those polled, it wanted clean air and was willing to pay for it. Second, and perhaps more important in the long run, is the fact that the basic problem the auto industry faces is not only that supplies of fossil fuels are dwindling, but also that burning a fossil fuel like gasoline, oil, or coal is inherently a very dirty process.

Given the consistent environmental problems that fossil fuels have caused us, one must wonder why we remain so committed to them. Don't alternatives exist—other combustible fuels that do not pose such a threat to our health? To a great extent the answer to this question is bound up with our historical dependence on a few large, highly successful corporations that have a natural tendency to shape our energy use patterns in a way that corresponds to their plans to maximize their profits. But such alternatives do exist. Their use on a mass scale is feasible, and they could meet our fuel needs in the near future, if we decided to escape from our dependence on fossil fuels.

Because most of these alternatives produce fewer toxic emissions when burned, they would pose far less of a health hazard than the emissions we now have to contend with. Methane, methanol, ethanol, hydrogen, and even the gases derived from burning wood or charcoal have all been used quite successfully in recent years to power motor vehicles and to generate electricity. In most cases these alternatives can be obtained from natural, domestic resources that are in little danger of depletion, at least in the United States. For example, methane, the major component of natural gas, can be derived from organic wastes, including manure and ordinary municipal garbage. Methanol and ethanol are alcohols, distillable through proven technologies from a variety of fermentable substances, including old newspapers (in the case of ethanol).[1] Hydrogen, contained in elemental form in every drop of water, is potentially the most abundant, least expensive, and cleanest of all these fuels, but it is also to date the most experimental.

Use of these clean alternatives on a mass national scale is no pipe dream. Brazil, which was importing some 1.2 million barrels of oil a day in 1980, has embarked on an aggressive campaign to convert virtually its entire transportation system to domestically produced ethanol by the early 1990s. Volkswagen, Ford, Fiat, GM, and Mercedes-Benz are taking the Brazilian government's commitment to this move so seriously that they have all begun to manufacture and market vehicles in Brazil designed and built to run on ethanol.

The one potential drawback to most of these comparatively clean fuels is that they contain carbon, and therefore give off carbon dioxide when burned. As we explained in Chapter 4, this process may have the potential to cause unprecedented and highly destructive changes to the entire biosphere.

In the process of photosynthesis plants naturally draw carbon dioxide out of the air and in turn give oxygen back, thus providing a simple, natural mechanism for preventing global carbon dioxide levels from building up to hazardous concentrations. But in the last few centuries man's activities have begun to overburden this natural control system. In the highly industrialized Northern Hemisphere the greatest source of carbon dioxide is the combustion of fossil fuels—coal, oil, natural gas, gasoline. In the less-industrialized Southern Hemisphere most carbon dioxide is caused by the combustion of wood for heating and cooking, though in some newly industrialized areas fossil fuel emissions are rapidly becoming much more of a problem than wood emissions.

Elevation of the earth's average atmospheric temperature by only a few degrees would, it is theorized, cause a global catastrophe, triggering a melting of polar ice caps that in turn would raise the level of the seas and oceans enough to flood millions of hectares of coastline. Changes in the weather patterns of the planet would be even more devastating, turning what are now our most fruitful croplands into deserts and bringing famine and forced migration to some of the most bountiful parts of the globe.

However, at present this is all still highly speculative. Though we do know that global carbon dioxide levels have slowly risen over the last century, it is not yet clear that this rise has been accompanied by a corresponding increase in global temperatures, and even if it has, the ultimate consequences of such a trend are still very much a matter of debate.

Given the risks of continued carbon dioxide buildup, there is only one truly clean alternative to fossil fuels, at least for powering motor vehicles. At the turn of the century electric cars were America's most popular form of motorized transport, and recent innovations in battery technology could make that true once again. Until a few years ago the major drawback to an electric car revival was the fact that conventional lead-acid batteries large enough to power an electric vehicle took hours to recharge, possessed an effective range of only fifty miles or so, and had a nasty habit of exploding if the hydrogen gas that collected in them wasn't regularly bled off. But the advent of zinc-chloride batteries has made twenty-minute recharge times and two hundred-mile ranges feasible. Thus cars equipped with engines costing only one-third as much as gasoline engines and producing no polluting emissions whatsoever are an imminent reality.

Health issues also arise in the other area of our lives that is strongly affected by rising oil prices—namely our slow, steady turn to coal as the best domestic alternative for generating electricity on a large scale. As we pointed out in Chapter 4, though burning coal in large generating plants has become a sophisticated and remarkably clean process, there are still serious environmental problems to be overcome. Nitrogen oxide emissions and acid rain are just two of them—there are others. For example, on a residential scale, if coal is used as a home heating fuel, its emissions must be as rigorously controlled as they are on an industrial scale if we are to avoid recurrences of the toxic coal smogs that made life in some industrial metropolises so unbearable and unhealthy in the first half of this century.

Even in more speculative areas, such as the production of "synfuels" (coal-derived substitutes for gas and liquid fuels), there are indications that we must, as always, proceed with great caution as we introduce new technologies into the biosphere. For example, researcher Barbara T. Walton of the Environmental Sciences Division of the Oak Ridge National Laboratory recently reported in *Science* that a substance called acridine, a waste compound produced in the creation of synfuels, causes teratogenic effects—extra eyes, antennae, heads—in young crickets hatched in soil treated with it.[2] As ominous as this sounds, it is too early to say just what it may imply about the risks of large-scale conversions to synfuel production to meet our fuel needs.

Despite their very real problems, the developed nations possess a great deal of human and material resources to use in their fight against environmental hazards. It is clear that it is in the underdeveloped and developing nations, which do not have such abundant technical and economic resources, that industrialization of any sort holds by far the worst potential for causing severe health problems. Here the picture of the future is not so promising. As in the developed nations, patterns of energy consumption will play an important role in determining how bad these problems become, but in most of the underdeveloped nations of the world energy factors will be far outweighed by the staggering demographics of continued rapid population growth and concentration in urban areas. By the end of this century the earth's population will have exploded from its present 4 billion to over 6 billion, and virtually all of this growth will have taken place in these countries.

According to one recent study the result will be the emergence of a dozen or so megalopolises—huge cities—each containing more than 10 million inhabitants, in parts of Asia, Latin America, Africa, and the Near East.[3] With few exceptions most of the populace in these vast urban sprawls will be destitute, malnourished, and existing in desperately crowded conditions. There will be little or no sanitation in many of these areas and inadequate supplies of fresh, pure drinking water. Under such conditions the greatest health threat will be the outbreak of infectious diseases such as typhus and typhoid fever. These diseases are best managed by alleviating the conditions of overcrowding and poor sanitation that give rise to them in the first place, but in these vast, impoverished cities such preventive measures are unlikely. The only recourse left to combat lethal epidemics will involve the mass application of chemicals to purify water and kill pests like rats and lice, a course of action that is not without serious risks.

But it will be the need to feed millions of new mouths that will play the strongest role in causing the increased use of synthetic chemicals—especially fertilizers and pesticides—in underdeveloped nations. Because very little additional arable acreage will become available anywhere in the world in the next twenty years, the only way to produce more food will be to increase yields on land that has already been brought under cultivation. To do so will inevitably mean, especially in poorer countries, heavier and heavier applications of these same compounds, with all the attendant problems we spelled out in Chapter 6.

Newer, safer pesticides—juvenile and aging hormones, sex attractants, and "safeners," substances added to food and water supplies to counter the toxic effects of pesticide residues—will be available from the developed nations, along with even more sophisticated agricultural tools, such as robot pesticide applicators and crop harvesters. But all this new, healthier technology will be costly, and few underdeveloped nations will be in a position to afford it. They will instead find themselves in a cruel economic bind: Desperate for food and lacking the land area to produce it themselves, they will have no choice but to rely on importing it in ever larger amounts, paying for it with credit obtained from those same food-exporting nations. But their mounting debts will in many cases preclude new credit arrangements that might finance the purchase of lifesaving advances in agricultural technology.

Genetic engineering may have some potential to avert this bleak picture. New, inexpensive strains of seeds, grains, and legumes have already been developed that show great promise. They not only contain more protein than traditional varieties, but also are able to resist insect depredation better and can in some cases draw much of their nitrogen out of the air, characteristics that obviously reduce their need for synthetic pesticides and fertilizers. However, the performance of bioengineered crops in field trials has to date been disappointing. Like many artificial organisms, they seem to lack some of the hardiness of their undomesticated cousins, and often prove susceptible to natural plant diseases.

This same bind between rapidly increasing industrialization and the inability to pay for the more sophisticated, more effective pollution control technology that industrialization requires is already causing serious environmental dislocations in many parts of the developing world. One often-cited Brazilian example, the newly industrialized city of Cubatão, is one of the most polluted places in the Southern Hemisphere. Its dozens of factories pour thousands of tons of toxic fumes a day into the city's air and millions of liters of toxic effluent into its four rivers. As a result there are no birds or insects in Cubatão's skies, no fish swim in those rivers, 25 percent of all medical emergencies involve significant respiratory distress, and the birth-defect rate is high, about double the 2 percent rate that is considered normal in developed nations for serious disorders. The technology to clean up Cubatão's problems already exists, but the costs of importing it are high enough to keep it from being put to use there.

Ironically, the same rush toward progress that has created Cubatão is also destroying the one natural resource that Brazil possesses in abundance, a resource that, if carefully managed, could provide the income to help control environmental hazards for centuries to come. In the Amazon Basin vast tracts of virgin rain forest are being opened up for commercial development and speculation for the first time in history. Unfortunately, the process is rapidly becoming an environmentally destructive one. Lured by bargain-basement prices and by the promise of high returns from crops, timber, and cattle raised on the cleared tracts, international cartels have ever since the late 1970s been steadily acquiring huge holdings in the Amazon Basin which they then either reforest in quick-cash crops like eucalyptus and pine or convert into grazing land for beef cattle. Already, more than three-quarters of the 5-million-square-kilometer Amazon jungle has been cleared and burned.

The price that has been paid for this accomplishment is a steep one; we may never know its true cost. Along with the disappearance of the trees that stood in the way of quick timber and grazing fodder have gone uncountable plant strains that may have held the biological keys to the secrets of new crop improvements as well as cures for many of the diseases that still plague mankind.

But the worst tragedy of all is that the clearing of the jungle has primarily been done by the time-honored practice of burning it where it stood or where it fell as it was cut or bulldozed. Fallen trees and accumulated debris were invariably burned rather than salvaged for their lumber. In the process of disposing of their debris in this manner, Amazon farmers and ranchers have made a significant contribution to what may become the foremost environmental threat of the next century, the problem of rising carbon dioxide levels that we discussed above.

In fact, the clearing of the Amazon has been a double blow. Since the trees and other vegetation that were cleared away were either not replaced at all or were replaced by species with one-tenth the capacity for transforming carbon dioxide into oxygen, the clearance program has also radically weakened nature's global photosynthetic system for keeping carbon dioxide levels in balance, at the same time that it was imposing unusually large demands on this system.

In this generally dark picture there is one bright exception—Mexico City. Paradoxically, it is the city that promises to become bigger than all others—by the year 2000 it will be populated by over 30 million people if present trends continue. Mexico City, however, is in a unique position. Blessed by fabulous wealth from its extensive oil reserves, sensitive to the fate that awaits the food-poor, over-populated nations of the future, Mexico City is already taking steps to avert some of its potential problems.

In 1979 a group of planners and government officials began the process of creating a master plan for the city's development, one that could forestall the problems posed by the sheer magnitude of Mexico City's growth. The plan envisions dividing the city into nine separate but interconnected satellites, each with its own residential, commercial, light industrial, and entertainment areas. To help abate Mexico City's severe air pollution problems, heavy manufacturing will be banished to sectors outside the city limits. Millions of trees will be planted to help combat the buildup of local carbon dioxide levels. If it can also find an effective way to deal with high levels of pollution from automobile emissions, Mexico City's plan has a good chance of working.

Despite the sharp differences that lie ahead for developed and underdeveloped nations we have seen over and over again that many of the substances of most concern to us, no matter where they are created, have no respect for national boundaries once they are released into the biosphere. The basic issue, as we are only now beginning to discover, is that these substances have the potential to affect us all, no matter where we live. In the near future, developed and underdeveloped nations may pursue different courses, meet different fates, but in the long run environmental health is a global issue, one that we must acknowledge and confront if we, as a species, are to continue to progress at all.

NOTES

Chapter 1

1. Lewis Thomas, *Lives of a Cell* (New York: Bantam, 1974), p. 122.
2. Samuel S. Epstein, M.D., *The Politics of Cancer* (Garden City, New York: Anchor Doubleday, 1979), pp. 62–63.
3. Rachel Carson, *Silent Spring* (New York: Fawcett Crest, 1962), p. 48.

Chapter 2

1. Carson, *Silent Spring,* pp. 42–43.
2. Thomas H. Maugh II, "Chemical Carcinogens: The Scientific Basis for Regulation," *Science,* Vol. 201 (September 29, 1978), p. 1200; and National Institute of Health, "Individual Differences in Cancer Susceptibility," *Annals of Internal Medicine,* Vol. 92, No. 6 (June 1980), p. 814.
3. John Higginson, M.D., "A Hazardous Society? Individual Versus Community Responsibility in Cancer Prevention," *American Journal of Public Health,* Vol. 66, No. 4 (April 1976), p. 364.

Chapter 3

1. I. J. Selikoff and D.H.K. Lee, *Asbestos and Disease* (New York: Academic Press, 1978), p. 22.
2. Waldemar C. Dreeson, M.D., et al., "A Study of Asbestosis in the Asbestos Textile Industry," Division of Industrial Hygiene, U.S. Public Health Service, Department of the Treasury (Washington, D.C., August 1938).

3. Philip M. Cook, Gary E. Glass, and James H. Tucker, "Asbestiform Amphibole Minerals: Detection and Measurement of High Concentrations in Municipal Water Supplies," *Science,* Vol. 185 (September 6, 1974), pp. 853–854.
4. Thomas J. Mason, Frank W. McKay, and Robert W. Miller, "Asbestos-Like Fibers in Duluth Water Supply, Relation to Cancer Mortality," *Journal of the American Medical Association,* Vol. 228, No. 8 (May 20, 1974), pp. 1019–1020; and B. S. Levy et al., "Investigating Possible Effects of Asbestos in City Water, Surveillance of Gastrointestinal Cancer Incidence in Duluth, Minnesota," *American Journal of Epidemiology,* Vol. 103 (1976), pp. 362–368.
5. Marty S. Kanarek et al., "Asbestos in Drinking Water and Cancer Incidence in the San Francisco Bay Area," *American Journal of Epidemiology,* Vol. 112, No. 1 (1980), pp. 54–72.
6. Raymond L. Murphy et al., "Floor Tile Installation as a Source of Asbestos Exposure," *American Review of Respiratory Diseases,* Vol. 104 (1971), pp. 576–580.
7. A. N. Rohl, A. M. Langer, and I. J. Selikoff, "Environmental Asbestos Pollution Related to Use of Quarried Serpentine Rock," *Science,* Vol. 197 (June 17, 1977), pp. 1319–1322.
8. Robert Sherrill, "Asbestos, the Saver of Lives, Has a Deadly Side," *The New York Times Magazine* (January 21, 1973), p. 64.
9. Selikoff, *Asbestos and Disease,* p. 173.
10. Carson, *Silent Spring,* p. 186.
11. Martin Kharrazi, Gad Potashnik, and John R. Goldsmith, "Reproductive Effects of Dibromochloropropane," *Israeli Journal of Medical Science,* Vol. 16, No. 5 (May 1980), pp. 403–406.
12. James R. Allen, Deborah A. Barsotti, and Laurine A. Carstens, "Residual Effects of Polychlorinated Biphenyls on Adult Nonhuman Primates and Their Offspring," *Journal of Toxicology and Environmental Health,* Vol. 6 (1980), pp. 55–66.
13. EPA, Office of Toxic Substances, *Support Document/Voluntary Environmental Impact Statement* (Washington, D.C.: EPA, April 1979), p. 22.
14. US Court of Appeals, District of Columbia, Petition No. 79-1580, Conclusion, pp. 40–41, argued June 6, 1980.
15. Thomas M. Wickizer, MPH, et al., "Polychlorinated Biphenyl Contamination of Nursing Mothers' Milk in Michigan," *American Journal of Public Health,* Vol. 71, No. 2 (February 1981), pp. 132–137.
16. EPA, Air and Hazardous Materials Division, Solid Waste Branch,

Region 6, "Incineration of PCBs: Summary of Approval Actions, Rollins Environmental Service and Energy Systems Company," 1201 Elm Street, Dallas, Texas: EPA, February 6, 1981, unpaginated.

Chapter 4

1. *San Francisco Chronicle,* October 9, 1980.
2. *Oakland Tribune,* October 11, 1980.
3. Bernardino Ramazzini, *De Morbis Artificum* (New York: Hafner, 1964), p. 72.
4. W. W. Kellog et al., "The Sulfur Cycle," *Science,* February 11, 1972, pp. 587–596.
5. Jonathan E. Ericson, Ph.D., Hiroshi Shirahata, M.S., and Clair C. Patterson, Ph.D., "Skeletal Concentrations of Lead in Ancient Peruvians," *The New England Journal of Medicine,* Vol. 300, No. 17 (April 26, 1979), pp. 946–951.
6. H. L. Needleman et al., "Deficits in Psychologic and Classroom Performance of Children With Elevated Dentine Lead Levels," *The New England Journal of Medicine,* Vol. 300, No. 13 (March 29, 1979), pp. 689–695.
7. Morris M. Joselow, Ph.D., et al., "Manganese Pollution in the City Environment and Its Relationship to Traffic Density," *American Journal of Public Health,* Vol. 68, No. 6 (June 1978), pp. 557–560.
8. Jack O. Hackney et al., "Experimental Studies on Human Health Effects of Air Pollution: II. Four-Hour Exposure to Ozone Alone and in Combination With Other Pollutant Gases," *Archives of Environmental Health,* Vol. 30 (August 1975), pp. 379–384.
9. Frederick Sweet et al., "Ozone Selectively Inhibits Growth of Human Cancer Cells," *Science,* Vol. 209 (August 22, 1980), pp. 931–933.
10. Winslow Fuller, "What's In the Air for Tightly Built Houses?" *Solar Age* (June 1981), p. 30.
11. Neville M. Lefcoe, M.D., FRCP (C) and Ion I. Inculet, M.E., "Particulates in Domestic Premises: II. Ambient Levels and Indoor-Outdoor Relationships," *Archives of Environmental Health,* Vol. 30 (December 1975), pp. 565–570.
12. Ralph E. Binder, M.D., MPH, et al., "Importance of the Indoor Environment in Air Pollution Exposure," *Archives of Environmental Health,* Vol. 31 (November/December 1976), pp. 277–279.

13. T. Kusuda et al., "Radioactivity as a Potential Factor in Building Ventilation," *ASHRAE Journal,* Vol. 21, No. 7 (July 1979), p. 33.
14. Takeshi Hirayama, M.D., "Non-smoking Wives of Heavy Smokers Have a Higher Risk of Lung Cancer: A Study From Japan," *British Medical Journal,* Vol. 282 (January 17, 1981), p. 184.
15. John Evelyn, *Fumifugium* (Short Title Catalogue) (London, 1661), p. 6.
16. Philip Drinker, "Deaths During the Severe Fog in London and Environs, Dec. 5 to 9, 1952," *AMA Archives of Industrial Hygiene and Occupational Medicine,* Vol. 7, No. 4 (April 1952), pp. 275–276.
17. WPD Logan, M.D., Ph.D., "Mortality in the London Fog Incident, 1952," *Lancet* (February 14, 1953), pp. 336–338.

Chapter 5

1. William M. Lewis and Michael C. Grant, "Acid Precipitation in the Western United States," *Science,* Vol. 207 (January 11, 1980), pp. 176–177.
2. W. Emile Coleman et al., "Identification of Organic Compounds in a Mutagenic Extract of a Surface Drinking Water by a Computerized Gas Chromatography/Mass Spectrometry System (GC/MS/COM)," *Environmental Science and Technology,* American Chemical Society, Vol. 14, No. 4 (May 1980), pp. 576–588.
3. Alice Hamilton, M.D., and Harriet L. Hardy, M.D., *Industrial Toxicology* (Acton, Massachusetts: Publishing Sciences Group, 3rd Edition, 1974), p. 143.
4. W. Eugene and Aileen M. Smith, *Minamata* (Holt, Rinehart & Winston, 1975).
5. Ian Rowland, Margaret Davies, Paul Grasso, "Biosynthesis of Methylmercury Compounds by the Intestinal Flora of the Rat," *Archives of Environmental Health* (January/February 1977), pp. 24–28.
6. W. P. Ridley, L. J. Dizikes, and J. M. Wood, "Biomethylation of Toxic Elements in the Environment," *Science,* Vol. 197, No. 4301 (July 22, 1977), pp. 329–332.
7. Smith, *Minamata,* p. 122.
8. Oiva I. Joensuu, "Fossil Fuels As a Source of Mercury Pollution," *Science,* Vol. 172 (June 4, 1971), pp. 1027–1028.
9. B. H. Olson and R. C. Cooper, "Comparison of Aerobic and

Anaerobic Methylation of Mercuric Chloride by San Francisco Bay Sediments," *Water Research,* Vol. 10 (1976), pp. 113–116.

10. EPA, "Acid Rain," Booklet # 600/9-79-036 (Washington, DC: EPA, July 1980), p. 17.
11. Ronald J. Kuzma, Cecilia M. Kuzma, and C. Ralph Buncher, "Ohio Drinking Water Source and Cancer Rates," *American Journal of Public Health,* Vol. 67, No. 8 (August 1977), pp. 725–729.
12. Robert W. Tuthill and Gary S. Moore, "Drinking Water Chlorination: A Practice Unrelated to Cancer Mortality," *Journal of the American Water Works Association,* Vol. 72, No. 10 (October 1980), pp. 570–583.
13. NAS, *Drinking Water and Health* (Washington, D.C.: NAS, 1977).
14. *The Federal Register,* Vol. 45, No. 231 (November 28, 1980), pp. 79318–79379.
15. Clean Water Act, PL 92-500 (Washington, D.C.: USGPO, 1977), p. 112.
16. Council of Environmental Quality, *Environmental Quality: The Tenth Annual Report of the Council on Environmental Quality* (Washington, D.C.: USGPO, 041-011-00047-5, December 1979).

Chapter 6

1. Hugh Trowell, M.D., *Diet Related to Killer Diseases,* Vol. IV, Hearings Before the Select Committee on Nutrition and Human Needs of the U.S. Senate, 95th Congress, 1st session (March 31, 1977), p. 57.
2. Bandura S. Reddy and Hideki Mori, "Effect of Dietary Wheat Bran and Dehydrated Citrus Fiber on 3,2-Dimethyl-4-Aminobiphenyl-Induced Intestinal Carcinogenesis in F344 Rats," *Carcinogenesis,* Vol. 2, No. 2 (1981), pp. 21–25.
3. B. E. Brush and J. K. Altland, "Goiter Prevention with Iodized Salt: Results of a Thirty-Year Study," *The Journal of Clinical Endocrinology,* Vol. 12 (1952), pp. 1380–88.
4. DHEW, *Report of the Secretary's Commission on Pesticides and Their Relationship to Environmental Health,* Washington, D.C.: GPO (December 1969), p. 211.
5. Carson, *Silent Spring,* p. 52.
6. W. F. Durham, J. F. Armstrong, and G. E. Quinby, "DDT and DDE Content of Complete Prepared Meals," *Archives of Environmental Health,* Vol. 11 (1965), pp. 641–647.
7. EPA, *Pilot National Environmental Profile: 1977* (Washington, D.C.: The Agency, October 1980), p. 50.

8. Peter Milius and Dan Morgan, "How a Toxic Pesticide Was Kept Off the Market," *The Washington Post,* Sunday, December 26, 1976.
9. Milius and Morgan, *loc. cit.*
10. Tom Curtis, "Danger: Men Working," *Texas Monthly* (May 1978), pp. 186–187.
11. M. C. Richardson, H. L. Richardson, and R. C. Calvert, "Morphological Lesions in the Sciatic Nerve of Chickens Paralyzed by Leptophos," unpublished abstract, meeting of the American Association of Pathologists and Bacteriologists (1976).
12. R. J. Kuhr and H. W. Dorough, *Carbamate Insecticides: Chemistry, Biochemistry, and Toxicology* (Cleveland, Ohio: CRC Press, 1976), p. 2.
13. DHEW, *op. cit.,* pp. 165–167.
14. John E. Casida, *Pyrethrum: The Natural Insecticide* (New York: Academic Press, 1973).
15. *Ibid.,* p. 132.
16. *Ibid.,* p. 136.
17. John E. Casida, "Pyrethrum Flowers and Pyrethroid Insecticides," *Environmental Health Perspectives,* Vol. 35 (February 1980), p. 132.
18. DHEW, *op. cit.,* p. 104.
19. Floyd M. Ashton and Alden S. Crafts, *Mode of Action of Herbicides* (New York: John Wiley and Sons, 1973), p. 5.
20. EPA, *Pilot National Environmental Profile: 1977* (Washington, D.C.: The Agency, October 1980), p. 48.
21. Julius J. Menn, "Contemporary Frontiers in Chemical Pesticide Research," *Journal of Agricultural and Food Chemistry,* Vol. 28, No. 1 (February 1980), p. 3.
22. CEQ and the Department of State, *The Global 2000 Report to the President: Entering the Twenty-First Century* (Washington, D.C.: USGPO, Vol. I, 1980), p. 16.
23. Ashton and Crafts, *Mode of Action of Herbicides,* p. 4.
24. Thomas H. Milby, M.D., E. Lee Husting, Ph.D., and M. Donald Whorton, M.D., "Potential Health Effects Associated With the Use of Phenoxy Herbicides" (Berkeley, California: Environmental Health Associates, July 1980), p. 42.
25. John G. Fuller, *The Poison That Fell From The Sky* (New York: Berkeley, 1979).
26. Fuller, *op. cit.,* p. 49.
27. Alessandro Di Domenico et al., "Accidental Release of 2,3,7, 8-Tetrachlorodibenzopidoxin (TCDD) at Seveso, Italy," Part II:

TCDD Distribution in the Soil Surface Layer, *Journal of Exotoxicology and Environmental Safety,* Vol. 14, No. 3 (September 1980), pp. 298–320.

28. G. Reggiani, "Acute Human Exposure to TCDD in Seveso, Italy," *Journal of Toxicology and Environmental Health,* Vol. 6, No. 1 (January 1980), pp. 41–42.
29. Reggiani, *op. cit.*
30. M. Eriksson et al., "Soft-Tissue Sarcomas and Exposure to Chemical Substances: A Case-Referent Study," *British Journal of Industrial Medicine,* Vol. 30 (1981), pp. 27–33.
31. Kuhr and Dorough, *Carbamate Insecticides: Chemistry, Biochemistry, and Toxicology,* p. 234.
32. EPA, Office of Pesticide Programs, "Ethylene Dibromide: Position Document 2/3" (undated), p. 62.

Chapter 7

1. EPA, Office of Water and Waste Management, "Solid Waste Facts" (Washington, D.C.: EPA, October 1979), p. 13.
2. EPA, Office of Water and Waste Management, "Everybody's Problem: Hazardous Waste" (Washington, D.C.: EPA, 1980), p. 15.
3. EPA, "Solid Waste Facts: A Statistical Handbook" (Washington, D.C.: EPA), p. 9.
4. D. G. Ackerman et al., "At-Sea Incineration of Herbicide Orange Onboard the *M/T Vulcanus* (Washington, D.C.: EPA-600/2-78-086, April 1978), p. 119.

Chapter 8

1. Atomic Bomb Casualty Commission, *Fact Book,* National Academy of Sciences-National Research Council/Japanese National Institute of Health (1969), p. 57.
2. Jan Langman, M.C., Ph.D., *Medical Embryology* (Baltimore, Maryland: Williams and Wilkins, 1969), p. 89.
3. Alice Hamilton and Harriet Hardy, *Industrial Toxicology* (Acton, Massachusetts: Publishing Sciences Group, 1974), p. 383.
4. *Loc.cit.*
5. Harrison S. Martland, M.D., "Occupational Poisoning in Manufacture of Luminous Watch Dials," *Journal of the American Medical Association,* Vol. 92, No. 6 (February 9, 1929), p. 468.
6. *Ibid.*
7. T. Kusuda et al., "Radioactivity as a Potential Factor in Build-

ing Ventilation," *ASHRAE Journal,* Vol, 21, No. 7 (July 1979), p. 32.
8. *Ibid,* p. 33.
9. Fred C. Shapiro, "A Reporter at Large: Nuclear Waste," *The New Yorker,* October 18, 1981, p. 63.
10. Genevieve M. Matanoski et al., "The Current Mortality Rates of Radiologists and Other Physician Specialists: Specific Causes of Death," *American Journal of Epidemiology,* Vol. 101, No. 3 (March 1975), pp. 199–210.
11. Jacob I. Fabrikant, "The BEIR-III Report: Origin of the Controversy," *American Journal of Radiology,* Vol. 136 (January 1981), p. 213.
12. Edward P. Radford, "Human Health Effects of Low Doses of Ionizing Radiation: The BEIR-III Controversy," *Radiation Research,* Vol. 84 (1980), p. 392.
13. *Report of the President's Commission on the Accident At Three-Mile Island* (Washington, D.C.: USGPO, 1979), p. 34.
14. Ernest J. Sternglass, "The Lethal Path of TMI Fallout," *The Nation,* Vol. 232, No. 9 (March 7, 1981), pp. 267–273.
15. UPI, 3/21/81.
16. Daniel Ford, "A Reporter at Large: Three-Mile Island," *The New Yorker,* April 13, 1981, p. 109.

Postscript

1. U.S. Army Natick Research and Development Command, *Enzymatic Hydrolysis of Cellulosic Wastes to Fermentable Sugars and the Production of Alcohol,* Pollution Abatement Division, Ford Sciences Laboratory (Natick, Massachusetts, January 19, 1976).
2. Barbara Walton, "A Chemical Impurity Produces Extra Compound Eyes and Heads in Crickets," *Science,* Vol. 212 (April 1981), pp. 51–53.
3. The Council on Environmental Quality and the State Department, *Global 2000* (Washington, D.C.: USGPO, 1980).

APPENDIX A

FOR MORE INFORMATION

IN GENERAL

For general information on environmental issues, especially those concerning air and water pollution, pesticides, and toxic wastes, contact the local EPA office for your region.

EPA Region 1
(Connecticut, Maine, Massachusetts, New Hampshire, Rhode Island, Vermont)
JFK Federal Building
Boston, Massachusetts 02203
617-223-7210

EPA Region 2
(New Jersey, New York, Puerto Rico, Virgin Islands)
26 Federal Plaza
New York, New York 10007
212-264-2525

EPA Region 3
(Delaware, Maryland, Pennsylvania, Virginia, West Virginia, District of Columbia)
6th and Walnut Streets
Philadelphia, Pennsylvania 19106
215-597-9814

EPA Region 4
(Alabama, Georgia, Florida, Mississippi, North Carolina, South Carolina, Tennessee, Kentucky)
345 Courtland Street, N.E.
Atlanta, Georgia 30308
404-881-4727

EPA Region 5
(Illinois, Indiana, Ohio, Michigan, Wisconsin, Minnesota)
230 S. Dearborn
Chicago, Illinois 60604
312-353-2000

EPA Region 6
(Arkansas, Louisiana, Oklahoma, Texas, New Mexico)
1201 Elm Street
Dallas, Texas 75270
214-767-2600

EPA Region 7
(Iowa, Kansas, Missouri, Nebraska)
1735 Baltimore Avenue
Kansas City, Missouri 64108
816-374-5493

EPA Region 8
(Colorado, Utah, Wyoming, Montana, North Dakota, South Dakota)
1860 Lincoln Street
Denver, Colorado 80203
303-837-3895

EPA Region 9
(Arizona, California, Nevada, Hawaii, Guam, American Samoa, Trust Territories of the Pacific)
215 Fremont Street
San Francisco, California 94105
415-556-2320

EPA Region 10
(Alaska, Idaho, Oregon, Washington)
1200 Sixth Avenue
Seattle, Washington 98101
206-442-1220

FOR PESTICIDE INFORMATION

Texas Technical University's Medical School operates an excellent, toll-free phone information service on pesticides, called the Pesticide Hazard Assessment Project. Funded by the EPA, the project provides information on the correct use of pesticides, general and technical information about pesticides and their risks, and emergency information in cases of suspected human poisonings. The project's number is 800-292-7664 within Texas and 800-531-7790 everywhere else. Their mailing address is Texas Pesticide Hazard Assessment Project, P.O. Box 914, San Benito, Texas 78586.

ENVIRONMENTAL HEALTH INFORMATION

The Department of Health and Human Services (DHHS) is equipped to respond to general questions about a wide range of environmental health issues, primarily through its National Institute of Environmental Health Sciences (NIEHS). Information on the findings of NIEHS research programs can be obtained from:

Hugh James Lee
Public Information Office
National Institute of Environmental Health Sciences
P.O. Box 12233
Research Triangle Park, North Carolina 27709

The DHHS also conducts a National Toxicology Program, which can furnish general and technical information on the results of tests of numerous environmental contaminants for toxic effects. For general information write to:

Steven S. D'Arazien
Public Information
National Toxicology Program
P.O. Box 12233
Research Triangle Park, North Carolina 27709
919-541-3991

For information on technical issues write or call:

Technical Information Section
National Toxicology Program
Landow Building—Room 3AO6
7910 Woodmont Avenue
Bethesda, Maryland 20205
301-496-1152

NONESTABLISHMENT SOURCES

For a view of environmental issues that often differs from that expressed by federal agencies, contact:

Environmental Defense Fund
1525 18th Street, N.W.
Washington, D.C. 20036

Natural Resources Defense Fund
1725 I Street, N.W.
Washington, D.C. 20006

AN INTERNATIONAL PERSPECTIVE

Some of the global implications of environmental issues, particularly as they affect relations between the developed nations and the Third World, can be gained from United Nations studies. Contact:

United Nations Environment Programme
New York Liaison Office
Room A-3608
United Nations Plaza, New York 10017

CANCER INFORMATION

If you have a question about cancer, try calling the National Cancer Institute's toll-free Cancer Information Service, which is equipped to respond to inquiries about the causes and the treatment of cancer. The numbers are:

Alabama: 1-800-292-6201
Alaska: 1-800-638-6070
California (from area codes 213, 714, and 805 only): 1-800-252-9066; from all other California area codes: 1-800-638-6694
Connecticut: 1-800-922-0824
Delaware: 1-800-523-3586
District of Columbia: 636-5700
Florida: 1-800-432-5953
Georgia: 1-800-327-7332
Hawaii (Oahu only): 524-1234; other Islands: Ask operator for Enterprise 6702
Illinois: 1-800-972-0586
Kentucky: 1-800-432-9321
Maine: 1-800-225-7034
Maryland: 1-800-492-1444
Massachusetts: 1-800-952-7420
Minnesota: 1-800-582-5262
New Hampshire: 1-800-225-7034
New Jersey (northern): 1-800-223-1000; (southern): 1-800-523-3586
New York State: 1-800-462-7255
New York City: 212-794-7982
North Carolina: 1-800-672-0943
North Dakota: 1-800-328-5188
Ohio: 1-800-282-6522
Pennsylvania: 1-800-822-3963
South Dakota: 1-800-328-5188
Texas: 1-800-392-2040
Vermont: 1-800-225-7034
Washington: 1-800-552-7212
Wisconsin: 1-800-362-8038

All other states: 1-800-638-6694

FOOD ADDITIVES

For questions about food additives contact the nearest FDA Consumer Affairs and Information Office:

FDA Region 1*
585 Commercial Street
Boston, Massachusetts 02109
617-223-5857

FDA Region 2
850 Third Avenue
Brooklyn, New York 11232
212-965-5754

599 Delaware Avenue
Buffalo, New York 14202
716-846-4483

20 Evergreen Place
East Orange, New Jersey 07018
201-645-3265

P.O. Box S4427, Old San Juan Station
San Juan, Puerto Rico 00905
809-753-4443

FDA Region 3
Room 900 U.S. Customhouse
2nd and Chestnut Streets
Philadelphia, Pennsylvania 19106
215-597-0837

Pittsburgh Resident Inspection Post
7 Parkway Center, Suite 645
Pittsburgh, Pennsylvania 15220
412-644-2858

900 Madison Avenue
Baltimore, Maryland 21201
301-962-3731

Falls Church Resident Inspection Post
701 W. Broad Street, Room 309
Falls Church, Virginia 22046
703-285-2578

FDA Region 4
1182 W. Peachtree Street, N.W.
Atlanta, Georgia 30309
404-881-7355

P.O. Box 118
Orlando, Florida 32802

6501 N.W. 36th Street, Suite 200
Miami, Florida 33166
305-855-0900

297 Plus Park Boulevard
Nashville, Tennessee 37217
615-251-7127

FDA Region 5
433 West Van Buren Street
1222 Main Post Office Building
Chicago, Illinois 60607
312-353-7126

1141 Central Parkway
Cincinnati, Ohio 45202
513-684-3501

Columbus Resident Inspection Post
New Federal Building
85 Marconi Boulevard, Room 231
Columbus, Ohio 43215
614-469-7353

Cleveland Resident Inspection Post
601 Rockwell Avenue, Room 463
Cleveland, Ohio 44114
216-522-4844

1560 East Jefferson Avenue
Detroit, Michigan 48207
313-226-6260

Indianapolis Resident Inspection Post
575 North Pennsylvania Street, Room 693
Indianapolis, Indiana 46204
317-269-6500

240 Hennepin Avenue
Minneapolis, Minnesota 55401
612-725-2121

Milwaukee Resident Inspection Post
615 E. Michigan Street
Milwaukee, Wisconsin 53202
414-291-3094

*See EPA listing, above, for states included in each federal region.

FDA Region 6
500 South Ervay, Suite 470-B
Dallas, Texas 75201
214-767-5433

4298 Elysian Fields Avenue
New Orleans, Louisiana 70122
504-589-2420

Houston Station
1440 N. Loop, Suite 250
Houston, Texas 77009
713-226-5581

San Antonio Resident Inspection Post
419 S. Main Street, Room 301
San Antonio, Texas 78204
512-229-6737

FDA Region 7
1009 Cherry Street
Kansas City, Missouri 64106
816-374-3817

St. Louis Station
Laclede's Landing
808 North Collins Street
St. Louis, Missouri 63102
314-425-5021

Omaha Resident Inspection Post
1619 Howard Street
Omaha, Nebraska 68102
402-221-4675

FDA Region 8
500 U.S. Customhouse
19th and California Streets
Denver, Colorado 80202
303-837-4915

FDA Region 9
50 United Nations Plaza, Room 524
San Francisco, California 94102
415-556-2682

1521 W. Pico Boulevard
Los Angeles, California 90015
213-688-4395

FDA Region 10
909 First Avenue
Federal Office Building, Room 5003
Seattle, Washington 98174
206-422-5258

FDA Headquarters
5600 Fishers Lane, Room 13-82
Rockville, Maryland 20857
301-443-4166

A lively and informative alternative view of food additives and good nutrition can be obtained from:

Nutrition Action
Center for Science in the Public Interest
1755 S Street, N.W.
Washington, D.C. 20009

RADIATION

For information on radiation, especially questions pertaining to commercial nuclear reactors and radwastes disposal, write or call the Nuclear Regulatory Commission at its headquarters in Washington, D.C., or at any one of its five regional offices:

Nuclear Regulatory Commission
1717 H Street, N.W.
Washington, D.C. 20555
301-492-7000

NRC Region 1
(Maine, Vermont, New Hampshire, Massachusetts, Connecticut, Rhode Island, New Jersey, New York, Pennsylvania, Maryland, Delaware)
Office of Inspection and Enforcement
631 Park Avenue
King of Prussia, Pennsylvania 19406
214-337-5000

NRC Region 2
(West Virginia, Virginia, Kentucky, North Carolina, South Carolina, Tennessee, Georgia, Alabama, Mississippi, Florida)
Office of Inspection and Enforcement
101 Marietta Street, Suite 3100
Atlanta, Georgia 30303
404-221-5505

NRC Region 3
(Michigan, Minnesota, Wisconsin, Ohio, Indiana, Illinois, Iowa, Missouri)
Office of Inspection and Enforcement
799 Roosevelt Road
Glen Ellyn, Illinois 60137
312-932-2500

NRC Region 4
(Louisiana, Arkansas, Oklahoma, Kansas, Nebraska, South Dakota, North Dakota, Montana, Idaho, Wyoming, Utah, Colorado, New Mexico, Texas)
Office of Inspection and Enforcement
611 Ryan Plaza Drive, Suite 1000
Arlington, Texas 76011
817-465-8100

NRC Region 5
(Washington, Oregon, California, Nevada, Arizona, Hawaii, Alaska)
Office of Inspection and Enforcement
1450 Maria Lane, Suite 210
Walnut Creek, California 94596
415-943-3809

For information on both ionizing and nonionizing radiation from electronic products (e.g., TVs, smoke detectors, sunlamps, microwave ovens) and from X rays as well as other health care uses of radioactivity, contact the FDA's Bureau of Radiological Health:

Director
Technical Information Staff (HFX025)
Bureau of Radiological Health
5600 Fishers Lane
Rockville, Maryland 20857
301-443-3434

A good, nonestablishment point of view about the hazards of nuclear power can be obtained from the:

Union of Concerned Scientists
1384 Massachusetts Avenue
Cambridge, Massachusetts 02238
617-547-5552

APPENDIX B

EMERGENCY ASSISTANCE

THE POISON CONTROL CENTERS

Reliable advice on the best emergency treatment for virtually any acute exposure to environmental health hazards can best be handled by one of the nation's five hundred Poison Control Centers (PCCs). Of the 7 million calls they receive each year from physicians, fire and police departments, and the public, 80 percent deal with poisonings of children under five years old, and most of those involve exposure to common household products. Some PCCs are also able to answer questions about chronic and occupational exposures to toxic substances.

General information about the PCC system can be obtained from:

Director
Division of Poison Control
Food and Drug Administration
5600 Fishers Lane
Rockville, Maryland 20857
301-443-6260

or

American Association of Poison Control Centers
Department P
225 Dickinson
San Diego, California 92103

THE POISON CONTROL DIRECTORY*

ALABAMA

State Coordinator
205-832-3194
Department of Public Health
Montgomery 36117

Anniston
205-235-5121
N.E. Alabama Regional Medical Center
400 E. 10th Street
36201

Birmingham
205-933-4050
800-292-6678
The Children's Hospital
1601 6th Avenue S.
35233

Dothan
205-793-8111
Southeast Alabama Medical Center
P.O. Drawer 6987
36301

Gadsden
205-492-8111
Baptist Memorial Hospital
1007 Goodyear Avenue
35903

Mobile
205-471-7100
University of South Alabama Medical Center
2451 Fillingim Street
33617

Opelika
205-749-3411
Ext. 258
Lee County Hospital
2000 Pepperill Parkway
36801

Tuskegee
205-727-8488
John A. Andrews Hospital Emergency Room
Tuskegee Institute
36088

ALASKA

State Coordinator
907-465-3100
Department of Health and Social Services
Juneau 99811

Anchorage
907-274-6535
Anchorage Poison Center
Providence Hospital
3200 Providence Drive
99504

*Provided by the Food and Drug Administration, Division of Poison Control.

ARIZONA

State Coordinator
602-626-6016
800-362-0101 — Arizona Poison Control System
College of Pharmacy
Tucson 85724

Flagstaff 602-774-5233	Flagstaff Hospital and Medical Center of N. Arizona 1215 N. Beaver Street 86001	Tucson 602-626-6016 800-362-0101 (In Arizona)	Arizona Regional Poison Center Arizona Health Sciences Center University of Arizona 85724
Phoenix 602-253-3334	St. Lukes Hospital and Medical Center 525 N. 18th Street 85006	Yuma 602-344-2000	Yuma Regional Medical Center Avenue A and 24th Street 85364

ARKANSAS

State Coordinator
501-661-6161 — University of Arkansas
Medical Science Campus
Little Rock 72201

El Dorado 501-863-2266	Warner Brown Hospital 460 W. Oak Street 71730	Helena 501-338-6411 Ext. 340	Helena Hospital Newman Drive 72342
Fort Smith 501-452-5100 Ext. 2401	St. Edward's Mercy Medical Center 7301 Rogers Avenue 72903	Little Rock 501-661-5544 800-482-8948	University of Arkansas Medical Center Slot 522 4301 W. Markham Street 72201
501-441-5011	Sparks Regional Medical Center 1311 S. Eye Street 72901	Osceola 501-563-7180	Osceola Memorial Hospital 611 Lee Avenue W. 72370
Harrison 501-741-6141 Ext. 275 or 276	Boone County Hospital Emergency Room 620 N. Willow Street 72601	Pine Bluff 501-535-6800 Ext. 4706	Jefferson Hospital 1515 W. 42nd Avenue 71601

CALIFORNIA

State Coordinator
916-322-4336
Department of Health and Welfare
Emergency Medical Services
Sacramento 95814

Fresno
209-445-1222
Central Valley Regional Poison Control Center
Fresno Community Hospital and Medical Center
P.O. Box 1232
Fresno and R Streets
93715

Los Angeles
213-664-2121
Thomas J. Fleming Memorial Center
Children's Hospital of Los Angeles
P.O. Box 54700
4650 Sunset Boulevard
90054

Oakland
415-428-3000
Children's Hospital Medical Center of Northern California
51st and Grove Streets
94609

Orange
714-634-5988
634-6011
University of California
Irvine Medical Center
101 City Drive S.
92688

Sacramento
916-453-3692
800-852-7221
Sacramento Medical Center
2315 Stockton Boulevard
95817

San Diego
714-294-6000
San Diego Poison Information Center
University Hospital
225 Dickinson Street
92103

San Francisco
415-666-2845
800-792-0720
San Francisco General Hospital
1001 Potrero Avenue
94102

San Jose
408-279-5112
Santa Clara Valley Medical Center
751 S. Bascom Avenue
95128

COLORADO

State Coordinator
303-320-8476
Department of Health
Emergency Medical Services
Denver 80220

Denver
303-629-1123
800-332-3073
Rocky Mountain Poison Center
Denver General Hospital
W. 8th Avenue and Cherokee Streets
80204

CONNECTICUT

State Coordinator
203-674-3456

University of Connecticut Health Center
Farmington 06032

Bridgeport
203-384-3566

Bridgeport Hospital
267 Grant Street
06602

203-576-5178

St. Vincent's Hospital
2820 Main Street
06602

Danbury
203-797-7300

Danbury Hospital
95 Locust Avenue
06810

Farmington
203-674-3456
1-800-845-7633

Connecticut Poison Control Center
University of Connecticut
Health Center
06032

Middletown
203-347-9471

Middlesex Memorial Hospital
28 Crescent Street
06457

New Haven
203-789-3469

The Hospital of St. Raphael
1450 Chapel Street
06511

203-436-1960

Yale-New Haven Hospital
Department of Pediatrics
789 Howard Avenue
06504

Norwalk
203-852-2160

Norwalk Hospital
24 Stevens Street
06852

Waterbury
203-574-6011

St. Mary's Hospital
56 Franklin Street
06702

DELAWARE

State Coordinator
302-655-3389

Wilmington Medical Center
Delaware Division
Wilmington 19801

Wilmington
302-655-3389

Wilmington Medical Center
Delaware Division
501 W. 14th Street
19899

DISTRICT OF COLUMBIA

State Coordinator
202-673-6741
202-673-6736

Department of Human Services
Washington, D.C. 20009

Washington 202-625-3333	National Capitol Poison Center Georgetown University Hospital 3800 Reservoir Road 20007

FLORIDA

State Coordinator
904-487-1566

Department of Health and
Emergency Medical Services
Tallahassee 32301

Apalachicola 904-653-8853	George E. Weems Memorial Hospital P.O. Box 610 Franklin Square 32320	Ft. Lauderdale 305-463-3131 Ext. 1511	Broward General Medical Center Emergency Department 1600 S. Andrews Avenue 33316
Bradenton 813-746-5111 Ext. 466	Manatee Memorial Hospital 206 Second Street E. 33505	Fort Myers 813-332-1111 Ext. 285	Lee Memorial Hospital 2776 Cleveland Avenue P.O. Drawer 2218 33902
Daytona Beach 904-258-2002	Halifax Hospital Emergency Department P.O. Box 1990 32014	Ft. Walton Beach 904-392-1111 Ext. 106	General Hospital of Ft. Walton Beach 1000 Mar-Walt Drive 32548

Gainesville 904-392-3740	Shands Teaching Hospital and Clinics University of Florida 32610
Inverness 904-726-2800	Citrus Memorial Hospital 502 Highland Boulevard 32650
Jacksonville 904-389-7751 Ext. 8315	St. Vincent's Medical Center Barrs Street and Johns Avenue 32204
Lakeland 813-686-4913	Lakeland General Hospital Lakeland Hills Boulevard P.O. Box 480 33802
Leesburg 904-787-7222 Ext. 381	Leesburg General Hospital 600 E. Dixie 32748
Melbourne 305-727-7000 Ext. 675	James E. Holmes Regional Medical Center 1350 S. Hickory Street 32901
Miami 305-325-6799	Jackson Memorial Hospital Attn: Pharmacy 1611 N.W. 12th Avenue 33136
Naples 813-262-7838	Naples Community Hospital 350 7th Street N. 33940
North Miami Beach 305-653-3333	Parkway General Hospital, Inc. 160 N.W. 170th Street 33169
Ocala 904-732-1111 Ext. 187	Munroe Memorial Hospital 140 S.E. Orange Street P.O. Box 6000 32670
Orlando 305-841-5222	Orlando Regional Medical Center Orange Memorial Division 1414 S. Kuhl Avenue 32806
Panama City 904-769-1511 Ext. 415 or 416	Bay Memorial Medical Center 600 N. MacArthur Avenue 32401
Pensacola 904-434-4611 800-874-1555 800-342-3222	Baptist Hospital 1000 W. Moreno Street (In Florida Only) 32501
Punta Gorda 813-637-2529	Medical Center Hospital 809 E. Marion Avenue 33950
Rockledge 305-636-2211 Ext. 168	Wuesthoff Memorial Hospital 110 Longwood Avenue 32955

FLORIDA

State Coordinator
904-487-1566

Department of Health and
Emergency Medical Services
Tallahassee 32301

St. Petersburg
813-821-5858

Bay Front Medical Center, Inc.
701 6th Street, S.
33701

Sarasota
813-953-1332

Memorial Hospital
1901 Arlington Street
33579

Tallahassee
904-599-5411

Tallahassee Regional Medical Center
1300 Miccosukee Road
32304

Tampa
813-251-6995

Tampa General Hospital
Davis Island
33606

Titusville
305-268-6111

Jess Parrish Memorial Hospital
951 N. Washington Avenue
P.O. Drawer W
32780

West Palm Beach
305-655-5511
Ext. 4250

Good Samaritan Hospital
Palm Beach Lakes Boulevard
33402

Winter Haven
813-299-9701

Winter Haven Hospital, Inc.
200 Avenue F, N.E.
33880

GEORGIA

State Coordinator
404-894-5170

Department of Human Resources
Emergency Health Section

Albany
912-883-1800
Ext. 4152

Phoebe Putney Memorial Hospital
417 Third Avenue
31705

Athens
405-543-5215

Athens General Hospital
797 Cobb Street
30601

Atlanta
404-588-4400
800-282-5846
404-525-3323
(Deaf)

Georgia Poison Control Center
Grady Memorial Hospital
80 Butler Street, S.E.
30303

Augusta
404-722-9011
Ext. 2440

University Hospital
1350 Walton Way
30902

Columbus
404-324-4711
Ext. 6431

The Medical Center
710 Center Street
31902

Macon
912-742-1122
Ext. 1146

Medical Center of Central Georgia
777 Hemlock Street
31201

Rome
404-291-2196

Floyd Hospital
P.O. Box 233
31061

Savannah
912-355-5228

Savannah Regional Poison Center
Department of Emergency Medicine
Memorial Medical Center
P.O. Box 23089
31403

Thomasville
912-226-4121
Ext. 169

John D. Archbold Memorial Hospital
900 Gordon Avenue
31792

Valdosta
912-333-1110

South Georgia Medical Center
P.O. Box 1727
31601

Waycross
912-283-3030

Memorial Hospital
410 Darling Avenue
31501

GUAM

State Coordinator
646-5801

Guam Memorial Hospital
P.O. Box AX
Agana 96910

Agana
344-9265
344-9354

Pharmacy Service, Box 7696
U.S. Naval Regional Medical Center
(GUAM)
FPO San Francisco, California 96630

HAWAII

State Coordinator
808-531-7776

Department of Health
Honolulu 96801

Honolulu
800-941-4411
1-800-362-3585

Kapiolani-Children's Medical Center
1319 Punahou Street
96826

IDAHO

State Coordinator
208-334-2241
1-800-632-8000

Department of Health and Welfare
Boise 83701

Boise
208-376-1211
Ext. 707

Idaho Poison Center
St. Alphonsus Hospital
1055 N. Curtis Road
83704

1-800-632-8000
(Statewide No.)

Idaho Falls
208-522-3600

Consolidated Hospitals Emergency Department
900 Memorial Drive
83401

Pocatello
208-232-2733
Ext. 244
800-632-9490
(Idaho)

St. Anthony's Hospital
650 North 7th Street
83201

ILLINOIS

State Coordinator
217-785-2080

Division of Emergency Medical Services and Highway Safety
Springfield 62761

Chicago
312-942-5969
800-942-5969

Rush Presbyterian-St. Luke's Medical Center
1753 W. Congress Parkway
60612

Peoria
309-672-2334
800-322-5330

St. Francis Hospital and Medical Center
530 N.E. Glen Oak Avenue
61637

Springfield
217-753-3330
800-252-2022

St. John's Hospital
800 E. Carpenter
62702

INDIANA

State Coordinator
317-633-0332

Indiana State Board of Health
Hazardous Products Section and Division of Drug Control
Indianapolis 46206

City / Phone	Hospital	City / Phone	Hospital
Anderson 317-646-5198	Community Hospital 1515 N. Madison Avenue 46012	Evansville 812-426-3405	Deaconess Hospital 600 Mary Street 47710
317-646-8251	St. John's Hickey Memorial Hospital 2015 Jackson Street 46014	812-426-8000	Welborn Memorial Baptist Hospital 401 S.E. 6th Street 47713
Angola 219-665-2141 Ext. 146	Cameron Memorial Hospital, Inc. 416 East Maumee Street 46703	Fort Wayne 219-458-2211	Lutheran Hospital 3024 Fairfield Avenue 46807
Columbus 812-376-5277	Bartholomew County Hospital 2400 East 17th Street 47201	219-484-6636 Ext. 6000	Parkview Memorial Hospital 220 Randalia Drive 46805
Crown Point 219-738-2100	St. Anthony Medical Center Main at Franciscan Road 46307	219-423-2614	St. Joseph's Hospital 700 Broadway 46802
East Chicago 219-392-1700 392-7203	St. Catherine's Hospital 4321 Fir Street 46312	Frankfort 317-659-4731	Clinton County Hospital 1300 S. Jackson Street 46041
Elkhart 219-294-2621 800-382-9697	Elkhart General Hospital 600 East Boulevard 46514	Gary 219-886-4710	Methodist Hospital of Gary, Inc. 600 Grant Street 46402

INDIANA

State Coordinator 317-633-0332	Indiana State Board of Health Hazardous Products Section and Division of Drug Control Indianapolis 46206

Goshen 219-533-2141	Goshen General Hospital 200 High Park Avenue 46526	317-423-6271	St. Elizabeth's Hospital 1501 Hartford Street 47904
Hammond 219-932-2300 931-4477	St. Margaret's Hospital 25 Douglas Street 46320	LaGrange 219-463-2144	LaGrange County Hospital Route #1 46761
Indianapolis 317-924-3521	Methodist Hospital of Indiana, Inc. 1604 N. Capitol Avenue 46202	LaPorte 219-326-1234	LaPorte Hospital, Inc. 1007 Lincolnway 46350
317-630-7351 800-382-9097	Indiana Poison Center 1001 W. Tenth Street 46202	Lebanon 317-482-2700 Ext. 241	Witham Memorial Hospital 1124 N. Lebanon Street 46052
Kendallville 219-347-1100	McCray Memorial Hospital Hospital Drive 46755	Madison 812-265-5211 Ext. 109	King's Daughter's Hospital 112 Presbyterian Avenue P.O. Box 447 47250
Kokomo 317-453-8444	Howard Community Hospital 3500 S. LaFountain Street 46901	Marion 317-662-4693	Marion General Hospital Wabash and Euclid Avenues 46952
Lafayette 317-447-6811	Lafayette Home Hospital 2400 South Street 47902		

Muncie
317-747-3241
Ball Memorial Hospital
2401 University Avenue
47303

Portland
219-726-7131
Jay County Hospital
505 W. Arch Street
47371

Richmond
317-962-7010
Reld Memorial Hospital
1401 Chester Boulevard
47374

Shelbyville
317-392-3211
Ext. 52
William S. Major Hospital
150 W. Washington Street
46176

South Bend
219-237-7264
St. Joseph's Hospital
811 E. Madison Street
46622

Terre Haute
812-238-7000
Ext. 7523
Union Hospital, Inc.
1606 N. Seventh Street
47804

Valparaiso
219-464-8611
Ext. 232
312
or 334
Porter Memorial Hospital
814 LaPorte Avenue
46383

Vincennes
812-885-3348
The Good Samaritan Hospital
520 S. 7th Street
47591

IOWA

State Coordinator
515-281-4964
Department of Health
Des Moines 50319

Des Moines
515-283-6254
1-800-362-2327
Blank Memorial Hospital
1200 Pleasant Street
50308

Dubuque
319-588-8050
Mercy Medical Center
Mercy Drive
52001

Fort Dodge
515-573-7211
515-573-3101
(Night)
Trinity Regional Hospital
Kenyon Road
50501

Iowa City
319-356-2922
800-272-6477
University of Iowa Hospital
Poison Information Center
52240

Waterloo
319-235-3893
Allen Memorial Hospital Emergency Room
1825 Logan Avenue
50703

KANSAS

State Coordinator
913-862-9360
Ext. 451

Kansas Department of Health and Environment
Bureau of Food and Drug
Forbes Field
Topeka 66620

Atchison
913-367-2131

Atchison Hospital
1301 N. 2nd Street
66002

Dodge City
316-225-9050
Ext. 381

Dodge City Regional Hospital
Ross and Avenue "A"
P.O. Box 1478
67801

Emporia
316-343-6800
Ext. 545

Newman Memorial Hospital
12th and Chestnut Streets
66801

Fort Riley
913-239-7777
239-7778

Irwin Army Hospital
Emergency Room
66442

Fort Scott
316-223-2200
316-223-0476
(Night)

Mercy Hospital
821 Burke Street
66701

Great Bend
316-792-2511
Ext. 115

Central Kansas Medical Center
3515 Broadway
67530

Hays
913-628-8251

Hadley Regional Medical Center
201 E. 7th Street
67601

Kansas City
913-588-6633

University of Kansas Medical Center
39th and Rainbow Boulevard
66103

Lawrence
913-843-3680
Ext. 162
or 163

Lawrence Memorial Hospital
325 Maine Street
66044

Parsons
316-421-4880
Ext. 320

Labette County Medical Center
S. 21st Street
67357

Salina
913-827-5591
Ext. 112

St. John's Hospital
139 N. Penn Street
67401

Topeka
913-354-6100

Stormont-Vail Regional Medical Center
10th and Washburn Streets
66606

Wichita
316-688-2222

Wesley Medical Center
550 N. Hillside Avenue
67214

KENTUCKY

State Coordinator
502-564-3970 — Department for Human Resources
Frankfort 40601

Ashland 606-324-2222	King's Daughter's Hospital 2201 Lexington Avenue 41101
Fort Thomas 606-292-3216 1-800-352-9900	St. Lukes Hospital 85 N. Grand Avenue 41075
Lexington 606-278-3411 Ext. 363	Central Baptist Hospital 1740 S. Limestone Street 40503
606-233-5320	Drug Information Center University of Kentucky Medical Center 40536
Louisville 502-589-8222 1-800-722-5725	Poison Control Center NKC, Inc. P.O. Box 35070 40232
Murray 502-753-7588	Murray-Calloway County Hospital 803 Popular 420701
Owensboro 502-926-3030 Ext. 180 or 186	Owensboro-Daviess County Hospital 811 Hospital Court 42301
Paducah 502-444-5100 Ext. 105 or 180	Western Baptist Hospital 2501 Kentucky Avenue 42001
Prestonburg 606-886-8511 Ext. 132 or 160	Poison Control Center Highlands Regional Medical Center 41653
South Williamson 606-237-1010	Appalachian Regional Hospitals Central Pharmaceutical Service 2000 Central Avenue 25661

LOUISIANA

State Coordinator
504-342-2600
Bureau of Emergency Medical Services
Baton Rouge 70801

Alexandria 318-487-8111 Ext. 231	Rapides General Hospital Emergency Department P.O. Box 7146 71301	Monroe 318-342-3008	School of Pharmacy Northeast Louisiana University 700 University Avenue 71209
Baton Rouge 504-928-6558	Doctors Hospital 2414 Bunker Hill Drive 70808	318-325-6454	St. Francis Hospital P.O. Box 1901 71301
Layfayette 318-234-7381	Our Lady of Lourdes Hospital P.O. Box 3827 611 St. Landry Street 70501	New Orleans 504-568-5222	Charity Hospital 1532 Tulane Avenue 70140
Lake Charles 318-478-6800	Lake Charles Memorial Hospital P.O. Drawer M 70601	Shreveport 318-425-1524	LSU Medical Center P.O. Box 33932 71130

MAINE

State Coordinator
207-871-2950
Maine Poison Control Center
Portland 04102

Portland 207-871-2950 800-442-6305	Maine Medical Center Emergency Division 22 Bramhall Street 04102

MARYLAND

State Coordinator
301-528-7604
Maryland Poison Center
University of Maryland School of Pharmacy
Baltimore 21201

Baltimore
301-528-7701
1-800-492-2414
Maryland Poison Center
University of Maryland School of Pharmacy
636 W. Lombard Street
21201

Cumberland
301-722-6677
Tri-State Poison Center
Sacred Heart Hospital
900 Seton Drive
21502

MASSACHUSETTS

State Coordinator
617-727-2700
Department of Public Health
Boston 02111

Boston
617-232-2120
1-800-682-9211
Massachusetts Poison Control System
300 Longwood Avenue
02115

MICHIGAN

State Coordinator
517-373-1406
Department of Public Health
Emergency Medical Services
Lansing 48909

Adrian
517-263-2412
Emma L. Bixby Hospital
818 Riverside Avenue
49221

Battle Creek
616-963-5521
Community Hospital
Pharmacy Department
183 West Street
49016

Ann Arbor
313-764-5102
University Hospital
1405 E. Ann Street
48104

Bay City
517-894-3131
Bay Medical Center
100 15th Street
48706

MICHIGAN

State Coordinator
517-373-1406

Department of Public Health
Emergency Medical Services
Lansing 48909

Berrien Center 616-471-7761	Berrien General Hospital Dean's Hill Road 49102
Coldwater 517-279-7935	Community Health Center of Branch County 274 E. Chicago Street 49036
Detroit 313-494-5711 800-572-1655 800-462-6642	Southeast Regional Poison Center Children's Hospital of Michigan 3901 Beaubien 48201
313-927-7000	Mount Carmel Mercy Hospital Pharmacy Department 6071 W. Outer Drive 48235
Eloise 313-722-3748 313-724-3000 (Night)	Wayne County General Hospital 30712 Michigan Avenue 48132
Flint 313-766-0111 800-572-5396	Hurley Hospital 6th Avenue and Begole 48502
Grand Rapids 616-774-6794	St. Mary's Hospital 201 Lafayette, S.E. 49503
800-442-4571 800-632-2727	Western Michigan Regional Poison Center 1840 Wealthy, S.E. 49506
Jackson 517-788-4816	W. A. Foote Memorial Hospital 205 N. East Street 49201
Kalamazoo 616-383-7070 1-800-632-4177 (Intra-Michigan Watts)	Midwest Poison Center Borgess Medical Center 1521 Gull Road 49001
616-383-6409 1-800-442-4112	Bronson Methodist Hospital 252 E. Lovell Street 49006
Lansing 517-372-5112 372-5113	St. Lawrence Hospital 1210 W. Saginaw Street 48914

Marquette
906-228-9440
1-800-562-9781
(N. Michigan)

Marquette General Hospital
420 W. Magnetic Drive
49855

Midland
517-631-8100

Midland Hospital
4005 Orchard Drive
48640

Petoskey
616-347-0555

Northern Michigan Hospitals, Inc.
416 Connable
49770

Pontiac
313-858-7373
858-7374

Poison Information Center
St. Joseph Mercy Hospital
900 S. Woodward Avenue
48053

Port Huron
313-987-5555
987-5000

Port Huron Hospital
1001 Kearney Street
48060

Saginaw
517-755-1111

Saginaw General Hospital
1447 N. Harrison
48602

Traverse City
616-947-6140

Poison Information Center
Munson Medical Center
Sixth Street
49684

MINNESOTA

State Coordinator
612-296-5281

State Department of Health
Minneapolis 55404

Brainerd
218-829-2861
Ext. 211

St. Joseph's Hospital
56401

Duluth
218-727-6636

St. Luke's Hospital
Emergency Department
915 E. First Street
55805

218-726-4500

St. Mary's Hospital
407 E. 3rd Street
55805

Edina
612-920-4400

Fairview-Southdale Hospital
6401 France Avenue, So.
55435

MINNESOTA

State Coordinator
612-296-5281

State Department of Health
Minneapolis 55404

Fergus Falls
218-736-5475
Lake Region Hospital
56537

Fridley
612-786-2200
Unity Hospital
550 Osborne Road
55432

Mankato
507-625-4031
Immanuel-St. Joseph's Hospital
325 Garden Boulevard
56001

Minneapolis
612-371-6402
Fairview Hospital
Outpatient Department
2312 S. 6th Street
55406

612-347-3141
Hennepin Poison Center
Hennepin County Medical Center
701 Park Avenue
55415

Morris
612-589-1313
Stevens County Memorial Hospital
56267

Rochester
507-285-5123
Ext. 517
Southeastern Minnesota Poison Control Center
St. Mary's Hospital
1216 Second Street, S.W.
55901

St. Cloud
612-251-2700
Ext. 221
St. Cloud Hospital
1406 6th Avenue, N.
56301

St. Paul
612-221-2301
Bethesda Lutheran Hospital
559 Capitol Boulevard
55103

612-228-3132
St. John's Hospital
403 Maria Avenue
55106

612-298-8402
United Hospitals, Inc.
300 Pleasant Avenue
55102

612-221-2113
St. Paul-Ramsey Hospital
640 Jackson Street
55101

Willmar
612-235-4543
Rice Memorial Hospital
402 W. 3rd Street
56201

Worthington
507-372-2941
Worthington Reg. Hospital
1016 6th Avenue
56187

MISSISSIPPI

State Coordinator
601-354-6660

State Board of Health
Jackson 39205

Biloxi 601-388-1919	Gulf Coast Community Hospital 4642 West Beach Boulevard 39531	601-354-6650	State Board of Health Bureau of Disease Control 39205
601-377-6555 377-6556	USAF Hospital Keesler Keesler Air Force Base 39534	601-354-7660	University Medical Center 2500 N. State Street 39216
Brandon 601-825-2811 Ext. 487 or 463	Rankin General Hospital Pharmacy Department 350 Crossgates Boulevard 39042	Laurel 601-649-4000 Ext. 207 218 220 or 248	Jones County Community Hospital Jefferson Street at 13th Avenue 39440
Columbia 601-736-6303 Ext. 217	Marion County General Hospital Sumrall Road 39429	Meridian 601-433-6211	Meridian Regional Hospital Highway 39, N. 39301
Greenwood 601-453-9751 Ext. 2633	Greenwood-Leflore Hospital River Road 38930	Pascagoula 601-938-5162	Singing River Hospital Emergency Room 2609 Denny Avenue 39567
Hattiesburg 601-264-4235	Forrest County General Hospital 400 S. 28th Avenue 39401	University 601-234-1522	University of Mississippi School of Pharmacy 38677
Jackson 601-982-0121 Ext. 2345	St. Dominic-Jackson Memorial Hospital 969 Lakeland Drive 39216		

MISSOURI

State Coordinator
314-751-2713

Missouri Division of Health
Jefferson City 65102

Cape Girardeau 314-651-6235	St. Francis Medical Center St. Francis Drive 63701	Kirksville 816-626-2121	Kirksville Osteopathic Health Center Box 949 #1 Osteopathy Avenue 63501
Columbia 314-882-8091	University of Missouri Medical Center 807 Stadium Road 65201	Poplar Bluff 314-785-7721	Lucy Lee Hospital 2620 N. Westwood Boulevard 63901
Hannibal 314-221-0414 Ext. 101	St. Elizabeth's Hospital c/o Pharmacy Department 109 Virginia Street 63401	Rolla 314-364-1322	Phelps County Memorial Hospital 1000 W. 10th Street 65401
Jefferson City 314-635-7141 Ext. 173	The Bureau of Emergency Medical Services Missouri Division of Health P.O. Box 570 65102	St. Joseph 816-271-7580 232-8481	Methodist Medical Center Seventh to Ninth on Faraon Street 64501
Joplin 417-781-2727 Ext. 2305	St. John's Medical Center 2727 McClelland Boulevard 64801	St. Louis 314-772-5200 1-800-392-9111 (MO)	Cardinal Glennon Memorial Hospital for Children 1465 S. Grand Avenue 63104
Kansas City 816-234-3000	Children's Mercy Hospital 24th at Gillham Road 46108	314-367-2034	St. Louis Children's Hospital 500 S. Kingshighway 63110

Springfield 417-831-9746 1-800-492-4824	Ozark Poison Center Lester E. Cox Medical Center 1423 N. Jefferson Street 65802	West Plains 417-256-9111 Ext. 258 or 259	West Plains Memorial Hospital 1103 Alaska Avenue 65775
417-885-2115	St. John's Regional Health Center 1235 E. Cherokee 65802		

MONTANA

State Coordinator
406-449-3895

Department of Health and Environmental Sciences
Helena 59620

Helena 1-800-525-5042	Montana Poison Control System Cogswell Building 59620

NEBRASKA

State Coordinator
402-471-2122

Department of Health
Lincoln 68502

Omaha 402-390-5400 800-642-9999 (NE residents) 800-228-9515 (Surrounding states)	Nebraska Regional Poison Center Children's Memorial Hospital 8301 Dodge 68114

NEVADA

State Coordinator
702-885-4750
Department of Human Resources
Carson City 89710

Las Vegas 702-385-1277	Southern Nevada Memorial Hospital 1800 W. Charleston Boulevard 89102	Reno 702-789-3013	St. Mary's Hospital 235 W. 6th 89503
702-732-4989	Sunrise Hospital Medical Center 3186 S. Maryland Parkway 89109	702-785-4129	Washoe Medical Center 77 Pringle Way 89502

NEW HAMPSHIRE

The following Center is the State Coordinator

Hanover 603-643-4000	New Hampshire Poison Center May Hitchcock Hospital Hanover 03755

NEW JERSEY

State Coordinator
609-292-5666
Department of Health
Accident Prevention and Poison Control Program
Trenton 08625

Atlantic City 609-344-4081 Ext. 2359	Atlantic City Medical Center 1925 Pacific Avenue 08401	Boonton 201-334-5000 Ext. 186 or 187	Riverside Hospital Powerville Road 07055
Belleville 201-751-1000 Ext. 781 782 or 783	Clara Maas Memorial Hospital 1A Franklin Avenue 07109	Bridgeton 609-451-6600	Bridgeton Hospital Irving Avenue 08302

Camden
609-795-5554

West Jersey Hospital
Evesham Avenue and Voorhees Turnpike
08104

Denville
201-627-3000
Ext. 6063

St. Clare's Hospital
Pocono Road
07834

East Orange
201-672-8400
Ext. 223

East Orange General Hospital
300 Central Avenue
07019

Elizabeth
201-527-5059

St. Elizabeth's Hospital
225 Williamson Street
07207

Englewood
201-894-3440

Englewood Hospital
350 Engle Street
07631

Flemington
201-782-2121
Ext. 369

Hunterdon Medical Center
Route #31
08822

Livingston
201-992-5161

St. Barnabas Medical Center
Old Short Hills Road
07039

Long Branch
201-222-2210

Monmouth Medical Center
Emergency Department
Dunbar and 2nd Avenue
07740

Montclair
201-746-6000
Ext. 234

Mountainside Hospital
Bay and Highland Avenues
07042

Mount Holly
609-267-7877

Burlington County Memorial Hospital
175 Madison Avenue
08060

Neptune
201-775-5500
800-822-9761

Jersey Shore Medical Center
Fitkin Hospital
1945 Corlies Avenue
07753

Newark
201-926-7240
926-7241
926-7242
926-7243

Newark Beth Israel Medical Center
201 Lyons Avenue
07112

New Brunswick
201-828-3000
Ext. 425
or 308

Middlesex General Hospital
180 Somerset Street
08903

201-745-8527

St. Peter's Medical Center
245 Easton Avenue
08903

Newton
201-383-2121
Ext. 270
271
or 273

Newton Memorial
Hospital Street
175 High Street
07860

NEW JERSEY

State Coordinator
609-292-5666

Department of Health
Accident Prevention and Poison Control Program
Trenton 08625

Orange 201-266-2120	Hospital Center at Orange Emergency Department 188 S. Essex Avenue 07051
Passaic 201-473-1000 Ext. 441	St. Mary's Hospital 211 Pennington Avenue 07055
Perth Amboy 201-442-3700 Ext. 2501	Perth Amboy General Hospital 530 New Brunswick Avenue 08861
Phillipsburg 201-859-1500 Ext. 280	Warren Hospital 185 Roseberry Street 08865
Point Pleasant 201-892-1100 Ext. 385	Point Pleasant Hospital Osborn Avenue and River Front 08742
Princeton 609-734-4554	The Medical Center at Princeton 253 Witherspoon Street 08540
Saddle Brook 201-368-6025	Saddle Brook General Hospital 300 Market Street 07662
Somers Point 609-653-3515	Shore Memorial Hospital Brighton and Sunny Avenues 08244
Somerville 201-725-4000 Ext. 431 432 or 433	Somerset Medical Center Rehill Avenue 08876
Summit 201-522-2232	Overlook Hospital 193 Morris Avenue 07901
Teaneck 201-833-3000	Holy Name Hospital 718 Teaneck Road 07666
Trenton 609-396-1077	Helene Fuld Medical Center 750 Brunswick Avenue 08638
Union 201-687-1900 Ext. 237	Memorial General Hospital 1000 Galloping Hill Road 07083
Wayne 201-942-6900 Ext. 224 225 or 226	Greater Paterson General Hospital 224 Hamburg Turnpike 07470

NEW MEXICO

The following Center is the State Coordinator

Albuquerque
505-843-2551
1-800-432-6866

New Mexico Poison, Drug Information and Medical Crisis Center University of New Mexico
Albuquerque 87131

NEW YORK

State Coordinator
518-474-3785

Department of Health
Albany 12237

Binghamton
607-723-8929

Southern Tier Poison Center
Binghamton General Hospital
13903

607-798-5231

Our Lady of Lourdes Memorial Hospital
169 Riverside Drive
13905

Buffalo
716-878-7000
878-7654
878-7655

Western New York Poison Center
Children's Hospital of Buffalo
219 Bryant Street
14222

Dunkirk
716-366-1111
Ext. 414
or 415

Brooks Memorial Hospital
10 W. 6th Street
14048

East Meadow
516-542-2323
542-2324
542-2325

Long Island Poison Center
Nassau County Medical Center
2201 Hempstead Turnpike
11554

Elmira
607-737-4100

Arnot Ogden Memorial Hospital
Roe Avenue and Grove Street
14901

607-734-2662

St. Joseph's Hospital Health Center
555 E. Market Street
14901

Endicott
607-754-7171

Ideal Hospital
600 High Street
13760

NEW YORK

State Coordinator
518-474-3785

Department of Health
Albany 12237

Glens Falls 518-792-3151 Ext. 456	Glens Falls Hospital 100 Park Street 12801
Jamestown 716-487-0141 484-8648	W.C.A. Hospital 207 Foote Avenue 14701
Johnson City 607-773-6611	Wilson Memorial Hospital 33 Harrison Street 13790
Kingston 914-331-3131	Kingston Hospital 396 Broadway 12401
New York City 212-340-4494 764-7667	NY City Poison Center Department of Health Bureau of Laboratories 455 First Avenue 10016
Nyack 914-358-6200 Ext. 451 or 452	Hudson Valley Poison Center Nyack Hospital N. Midland Avenue 10960
Rochester 716-275-5151	Finger Lakes Poison Center LIFE LINE University of Rochester Medical Center 14620
Schenectady 518-382-4039 382-4121	Ellis Hospital 1101 Nott Street 12308
Syracuse 315-476-7529 473-5831	Syracuse Poison Information Center 750 E. Adams Street 13210
Troy 518-272-5792	St. Mary's Hospital 1300 Massachusetts Avenue 12180
Utica 315-798-6200 798-6223	St. Luke's Hospital Center P.O. Box 479 13502
Watertown 315-788-8700	House of the Good Samaritan Hospital Washington and Pratt Streets 13602

NORTH CAROLINA

State Coordinator
919-684-8111
Duke University Medical Center
Durham 27710

Asheville
704-255-4490
Western North Carolina
Poison Control Center
509 Biltimore Avenue
28801

Charlotte
704-379-5827
Mercy Hospital
2001 Vail Avenue
28207

Greensboro
919-379-4105
Moses Cone Hospital
1200 N. Elm Street
27420

Hendersonville
704-693-6522
Ext. 555
556
Margaret R. Pardee Memorial Hospital
Fleming Street
28739

Hickory
704-322-6649
Catawba Memorial Hospital
Fairgrove-Church Road
28601

Jacksonville
919-353-7610
Onslow Memorial Hospital
Western Boulevard
28540

Wilmington
919-343-7046
New Hanover Memorial Hospital
2131 S. 17th Street
28401

NORTH DAKOTA

State Coordinator
701-224-2388
Department of Health
Bismarck 58505

Bismarck
701-223-4357
Bismarck Hospital Emergency Dept.
300 N. 7th Street
58501

Fargo
701-280-5575
St. Luke's Hospital
Fifth Street at Mills Avenue
58122

NORTH DAKOTA

State Coordinator
701-224-2388
Department of Health
Bismarck 58505

Grand Forks
701-780-5000
United Hospital
1200 S. Columbia Road
58201

Minot
701-857-2553
St. Joseph's Hospital
Third Sreet and Fourth Avenue, S.E.
58701

Williston
701-572-7661
Mercy Hospital
1301 15th Avenue, W.
58801

OHIO

State Coordinator
614-466-5190
Department of Health
Columbus 43216

Akron
216-379-8562
1-800-362-9922
Children's Hospital Medical Center of Akron
281 Locust Street
44308

Canton
216-452-9911
Ext. 203
Aultman Hospital Emergency Room
2600 Sixth Street, S.W.
44710

Cincinnati
513-872-5111
Drug and Poison Information Center
Bridge Medical Science Building
Room 7701
231 Bethesda Avenue
45267

Cleveland
216-231-4455
Academy of Medicine
11001 Cedar Avenue
44106

Columbus
614-228-1323
Ohio Poison Center
Children's Hospital
700 Children's Drive
43205

Dayton
513-222-2227
Children's Medical Center
One Children's Plaza
45404

Lorain
216-282-2220

Lorain Community Hospital
3700 Kolbe Road
44053

Mansfield
419-522-3411
Ext. 545

Mansfield General Hospital
335 Glessner Avenue
44903

Springfield
513-325-1255

Community Hospital
2615 E. High Street
44505

Toledo
419-381-3897

Poison Information Center
Medical College Hospital
P.O. Box 6190
43679

Youngstown
216-746-2222

Mahoning Valley Poison Center
1044 Belmont Avenue
44501

Zanesville
614-454-4221

Poison Information Center
Bethesda Hospital
2951 Maple Avenue
43701

OKLAHOMA

State Coordinator
405-271-5454
800-522-4611

Oklahoma Poison Control Center
Oklahoma Children's Memorial Hospital
P.O. Box 26307
Oklahoma City 73126

Ada
405-322-2323
Ext. 200

Valley View Hospital
1300 E. 6th Street
74820

Ardmore
405-223-5400

Memorial Hospital of Southern Oklahoma
1011-14th Avenue
73401

Lawton
405-355-8620

Comanche County Memorial Hospital
3401 Gore Boulevard
73501

McAlester
918-426-1800
Ext. 240

McAlester General Hospital Inc., W.
P.O. Box 669
74501

OKLAHOMA

State Coordinator
405-271-5454
800-522-4611

Oklahoma Poison Control Center
Oklahoma Children's Memorial Hospital
P.O. Box 26307
Oklahoma City 73126

Oklahoma City
405-271-5454
800-522-4611

Oklahoma Poison Control Center
Oklahoma Children's Memorial Hospital
P.O. Box 26307
73126

Tulsa
918-584-1351
Ext. 6165

Hillcrest Medical Center
1653 East 12th
74104

Ponca City
405-765-3321

St. Joseph Medical Center
14th and Hartford
74601

OREGON

The following Center is the State Coordinator

Portland
503-225-8968
1-800-452-7165

Oregon Poison Control and Drug Information Center
University of Oregon Health Sciences Center
Portland 97201

PANAMA

Panama City
52-7105

U.S.A. MEDDAC Panama
Gorgas U.S. Army Hospital
APO Miami 34004

PENNSYLVANIA

State Coordinator
717-787-2307

Director, Division of Epidemiology
Department of Health
P.O. Box 90
Harrisburg 17108

Allentown 215-433-2311	Lehigh Valley Poison Center 17th and Chew Streets 18102
Altoona 814-946-3711	Altoona Region Poison Center 2500 Seventh Avenue 16603
Bloomsburg 717-784-7121	The Bloomsburg Hospital 549 Fair Street 17815
Bradford 814-368-4143	Bradford Hospital Interstate Parkway 16701
Bryn Mawr 215-896-3577	The Bryn Mawr Hospital 19010
Chester 215-494-0721 Ext. 232	Sacred Heart General Hospital 9th and Wilson Street 19013
Clearfield 814-765-5341	Clearfield Hospital 809 Turnpike Avenue 16830
Coaldale 717-645-2131	Coaldale State General Hospital 18218
Coudersport 814-274-9300	Charles Cole Memorial Hospital RD #3 Route 6 16915
Danville 717-275-6116	Susquehanna Poison Center Geisinger Medical Center North Academy Avenue 17821
Doylestown 215-345-2283	Doylestown Hospital 595 W. State Street 18901
East Stroudsburg 717-421-4000	Pocono Hospital 206 E. Brown Street 18301
Easton 215-258-6221	Easton Hospital 21st and Lehigh Street 18042

State Coordinator
717-787-230

PENNSYLVANIA
Director, Division of Epidemiology
Department of Health
P.O. Box 90
Harrisburg 17108

Erie
814-454-2120

Doctors Osteopathic Hospital
252 W. 11th Street
16501

814-864-4031

Erie Osteopathic Hospital
5515 Peach Street
16509

814-452-4242

Hamot Medical Center
201 State Street
16512

814-452-3232

Northwest Poison Center
St. Vincent's Health Center
P.O. Box 740
16512

Gettysburg
717-334-2121

Annie M. Warner Hospital
S. Washington Street
17325

Greensburg
412-832-4000

Westmoreland Hospital Association
532 W. Pittsburgh Street
15601

Hanover
717-637-3711

Hanover General Hospital
300 Highland Avenue
17331

Harrisburg
717-782-3639

Harrisburg Hospital
S. Front and Mulberry Street
17101

717-782-4141
Ext. 4132

Polyclinic Hospital
3rd and Polyclinic Avenue
17105

Hershey
717-634-6111
534-8955

Capital Area Poison Center
The Milton S. Hershey Medical Center
University Drive
17033

Jeannete
412-527-3551

Jeannete District Memorial Hospital
600 Jefferson Avenue
15644

Jersey Shore
717-398-0100

Jersey Shore Hospital
Thompson Street
17740

Johnstown 814-535-5351	Conemaugh Valley Memorial Hospital 1086 Franklin Street 15905
814-535-5352	Cambria-Somerset Poison Center Lee Hospital 320 Main Street 15901
814-535-5353	Mercy Hospital 1020 Franklin Street 15905
Lancaster 717-299-5511	Lancaster General Hospital 555 N. Duke Street 17604
717-299-4546	St. Joseph's Hospital 250 College Avenue 17604
Lansdale 215-368-2100	North Penn Hospital 7th and Broad Street 19446
Lebanon 717-272-7611	Good Samaritan Hospital 4th and Walnut Streets 17042
Lehighton 215-377-1300	Gnaden-Huetten Memorial Hospital 11th and Hamilton Street 18235
Lewiston 717-248-5411	Lewistown Hospital Highland Avenue 17044
Muncy 717-546-8282	Muncy Valley Hospital P.O. Box 340 17756
Nanticoke 717-735-5000	Nanticoke State Hospital N. Washington Street 18634
Paoli 215-648-1043	Paoli Memorial Hospital 19301
Philadelphia 215-922-5523 215-922-5524	Philadelphia Poison Information 321 University Avenue 19104
Philipsburg 814-342-3320	Philipsburg State General Hospital 16866
Pittsburgh 412-681-6669	Children's Hospital 125 Desoto Street 15213
Pittston 717-654-3341	Pittston Hospital Oregon Heights 18640

Pottstown
215-327-7000

Pottstown Memorial Medical Center
High Street and Firestone Boulevard
19464

Pottsville
717-622-3400
Ext. 270

Good Samaritan Hospital
E. Norwegian and Tremont Streets
17901

Reading
215-376-4881
Ext. 267

Community General Hospital
145 N. 6th Street
19601

Sayre
717-888-6666

The Robert Packer Hospital
Guthrie Square
18840

Sellersville
215-257-3611

Grandview Hospital
18960

Somerset
814-443-2626

Somerset Community Hospital
225 S. Center Avenue
15501

State College
814-238-4351

Centre Community Hospital
16801

Titusville
814-827-1851

Titusville Hospital
406 W. Oak Street
16354

Tunkhannock
717-836-2161

Tyler Memorial Hospital
RD #1
18657

York
717-843-8623

Memorial Osteopathic Hospital
325 S. Belmont Street
17403

717-771-2311

York Hospital
1001 S. George Street
17405

PENNSYLVANIA

State
Coordinator
717-787-2307

Director, Division of Epidemiology
Department of Health
P.O. Box 90
Harrisburg 17108

PUERTO RICO

State Coordinator
809-765-4880
809-765-0615 — University of Puerto Rico
Rio Piedras

Arecibo 809-878-6467 (Information only)	District Hospital of Arecibo 00613	Ponce 809-842-8364	District Hospital of Ponce 00731
Fajardo 809-863-0939 Ext. 202 or 203	District Hospital of Fajardo 00649	Rio Piedras 809-754-8535 Ext. 8536 8537 or 8538	Children's Hospital Center of Puerto Rico 00936
Mayaguez 809-832-8686 Ext. 1224	Mayaguez Medical Center Department of Health P.O. Box 1868 00709	San Juan 809-753-4849 (Information only)	Pharmacy School Medical Science Campus 00936

RHODE ISLAND

The following Center is the State Coordinator

Providence 401-277-5727	Rhode Island Poison Control Center Rhode Island Hospital Providence 02902

SOUTH CAROLINA

State Coordinator
803-758-5654 — Department of Health and Environmental Control
Columbia 29201

Charleston 803-792-4201 800-845-7633 800-922-0193	National Pesticide Telecommunications Network Medical University of South Carolina 171 Ashley Avenue 29403	Columbia 803-765-7359 1-800-922-1117	Palmetto Poison Center University of South Carolina College of Pharmacy 29208

SOUTH DAKOTA

State Coordinator
605-773-3361
Department of Health
Pierre 57501

Aberdeen
605-225-1880
1-800-592-1889
The Dakota Midland Poison Control Center
57401

Rapid City
605-341-8222
1-800-742-8925
Rapid City Regional Hospital Main
353 Fairmont Boulevard
57701

Sioux Falls
605-336-3894
1-800-952-0123
McKennan Hospital Poison Center
800 East 21st Street
57101

TENNESSEE

State Coordinator
615-741-2407
Department of Public Health
Division of Emergency Services
Nashville 37216

Chattanooga
615-755-6100
T. C. Thompson Children's Hospital
910 Blackford Street
37403

Columbia
615-381-4500
Maury County Hospital
1224 Trotwood Avenue
38401

Cookeville
615-526-4818
Cookeville General Hospital
142 W. 5th Street
38501

Jackson
901-424-0424
Madison General Hospital
708 W. Forest
38301

Johnson City
615-461-6111
Memorial Hospital
Boone and Fairview Avenue
37601

Knoxville
615-971-3261
Memorial Research Center and Hospital
1924 Alcoa Highway
37920

Memphis
901-528-6048
Southern Poison Center
University of Tennessee
College Pharmacy
26 Dunlap Street
38163

Nashville
615-322-3391
Vanderbilt University Hospital
21st and Garland
37232

TEXAS

State Coordinator
512-458-7254

Texas Department of Health
Division of Occupational Health
Austin 78756

Abilene
915-677-7762

Hendrick Hospital
19th and Hickory Streets
79601

Amarillo
806-376-4292

Amarillo Emergency Receiving Center
Amarillo Hospital District
P.O. Box 1110
2103 W. 6th Street
79106

Austin
512-478-4490
476-6461

Brackenridge Hospital
14th and Sabine Streets
78701

Beaumont
713-833-7409

Baptist Hospital of Southeast Texas
P.O. Box 1591
College and 11th Street
77701

Corpus Christi
512-881-4559

Memorial Medical Center
P.O. Box 5280
2606 Hospital Boulevard
78405

El Paso
915-533-1244

R.E. Thomason General Hospital
P.O. Box 20009
4815 Alameda Avenue
79905

Fort Worth
817-927-2007

W. I. Cooke Children's Hospital
1212 Lancaster
76102

Galveston
713-765-1420

Southeast Texas Poison Center
8th and Mechanic Streets
77550

Harlingen
512-421-1860
421-1859

Valley Baptist Hospital
P.O. Box 2588
2101 S. Commerce Street
78550

Houston
713-654-1701

Southeast Texas Poison Control Center
8th and Mechanic Streets
77550

Laredo
512-724-6247

Mercy Hospital
1515 Logan Street
78040

Lubbock
806-792-1011

Methodist Hospital Pharmacy
3615 19th Street
79410

Midland
915-685-1111

Midland Memorial Hospital
1908 W. Wall
79701

TEXAS

State Coordinator
512-458-7254
Texas Department of Health
Div. of Occupational Health
Austin 78756

Odessa
915-333-7111
Medical Center Hospital
P.O. Box 633
79760

Plainview
806-296-9601
Central Plains Regional Hospital
2601 Dimmitt Road
79072

San Angelo
915-653-6741
Ext. 210
Shannon West Texas Memorial Hospital
P.O. Box 1879
9 S. Magdalen Street
76901

San Antonio
512-223-6361
Ext. 295
Department of Pediatrics
University of Texas Health Science Center at San Antonio
7703 Floyd Curl Drive
78284

Tyler
214-597-0351
Medical Center Hospital
1000 S. Beckham Street
75701

Waco
817-756-8611
Hillcrest Baptist Hospital
3000 Herring Avenue
76708

Wichita Falls
817-322-6771
Wichita General Hospital
1600 8th Street
76301

UTAH

State Coordinator
801-533-6161
Utah Department of Health
Division Family Health Services
Salt Lake City 84113

Salt Lake City
801-581-2151
Intermountain Regional Poison Control Center
50 N. Medical Drive
84132

VERMONT

State Coordinator
802-862-5701

Department of Health
Burlington 05401

Burlington 802-658-3456	Vermont Poison Center Medical Center Hospital 05401

VIRGINIA

State Coordinator
804-786-5188

Bureau of Emergency Medical Services
Richmond 23219

Alexandria 703-379-3070	Alexandria Hospital 4320 Seminary Road 22314
Arlington 703-558-6161	Arlington Hospital 5129 N. 16th Street 22205
Blacksburg 703-951-1111	Montgomery County Community Hospital Route 460, S. 24060
Charlottesville 804-924-5543 1-800-446-9876 (Deaf out-of-state) 1-800-552-3723 (Deaf VA only)	Blue Ridge Poison Center University of Virginia Hospital 22903
Danville 804-799-2100 Ext. 3869	Danville Memorial Hospital 142 S. Main Street 22201
Falls Church 703-698-3600 698-3111	Fairfax Hospital 3300 Gallows Road 22046
Hampton 804-727-1131	Hampton General Hospital 3120 Victoria Boulevard 23661
Harrisonburg 703-433-9706	Rockingham Memorial Hospital 738 S. Mason Street 22801
Lexington 703-463-9141	Stonewall Jackson Hospital 22043

VIRGINIA

State Coordinator
804-786-5188

Bureau of Emergency Medical Services
Richmond 23219

Lynchburg
804-528-2066
Lynchburg General Marshall Lodge Hospital, Inc.
Tate Springs Road
24504

Nassawadox
804-442-8700
Northampton-Accomack Memorial Hospital
23413

Newport News
804-599-2050
Riverside Hospital
500 J. Clyde Morris Boulevard
23601

Norfolk
804-489-5288
DePaul Hospital
Granby Street at Kingsley Lane
23505

Petersburg
804-732-7220
Petersburg General Hospital
Mt. Erin and Adams Streets
23803

Portsmouth
804-398-5898
U.S. Naval Hospital
23708

Richmond
804-786-9123
Central Virginia Poison Center
Medical College of Virginia
Box 763 MCV Station
23298

Roanoke
703-981-7336
Roanoke Memorial Hospital
Belleview at Jefferson Street
P.O. Box 13367
24033

Staunton
703-885-6848
King's Daughter's Hospital
P.O. Box 2007
24401

Waynesboro
703-942-4096
Waynesboro Community Hospital
501 Oak Avenue
22980

Williamsburg
804-253-6005
Williamsburg Community Hospital
1238 Mt. Vernon Avenue
Drawer H
23185

VIRGIN ISLANDS

State Coordinator
809-774-1321
Ext. 275

Department of Health
St. Thomas 00801

St. Croix
809-773-1212
773-1311
Ext. 221

Charles Harwood Memorial Hospital
Christiansted
00820

809-772-0260
772-0212

Ingeborg Nesbitt Clinic
Fredericksted
00840

St. John
809-776-1469

Morris F. DeCastro Clinic
Cruz Bay
00830

St. Thomas
809-774-1321
Ext. 224
or 225

Knud-Hansen Memorial Hospital
00801

WASHINGTON

State Coordinator
206-522-7478

Department of Social and Health Services
Seattle 98115

Seattle
206-634-5252

Children's Orthopedic-Hospital and Medical Center
4800 Sandpoint Way, N.E.
98105

Spokane
509-747-1077
1-800-572-5842

Deaconess Hospital
W. 800 5th Avenue
99210

Tacoma
206-272-1281
Ext. 259

Mary Bridge Children's Hospital
South L Street
98405

Yakima
509-248-4400
1-800-572-9176

Central Washington Poison Center
Yakima Valley Memorial Hospital
2811 Tieton Drive
98902

WEST VIRGINIA

State Coordinator
304-348-2971
Department of Health
Charleston 25305

Charleston
1-800-642-3625
1-304-348-4211
West Virginia Poison System
3110 MacCorkle Avenue, S.E.
29208

WISCONSIN

State Coordinator
608-267-7174
Department of Health and Social Services
Division of Health
Madison 53701

Eau Claire
715-835-1515
Luther Hospital
1225 Whipple
54701

Green Bay
414-433-8100
Green Bay Poison Control Center
St. Vincent's Hospital
835 S. Van Buren Street
54305

LaCrosse
608-784-3971
St. Francis Hospital
700 West Avenue N.
54601

Madison
608-262-3702
Madison Area Poison Center
University Hospital and Clinic
600 Highland Avenue
53792

Milwaukee
414-931-4114
Milwaukee Children's Hospital
1700 W. Wisconsin
53233

WYOMING

State Coordinator
307-777-7955
Office of Emergency Medical Services
Department of Health and Social Services
Cheyenne 82001

Cheyenne
307-635-9256
Wyoming Poison Center
DePaul Hospital
2600 E. 18th Street
82001

FURTHER READING

Eater's Digest, Michael F. Jackson (Garden City, New York: Anchor Doubleday, 1972).

Laying Waste, Michael Brown (New York: Washington Square Press, 1979).

Minamata, W. Eugene Smith and Aileen M. Smith (New York: Holt, Rinehart & Winston, 1975).

Radiation and Life, Eric Hall (New York: Pergamon Press, 1976).

Silent Spring, Rachel Carson (Boston: Houghton Mifflin, 1962).

INDEX